1968

ANALYTICAL TRIGONOMETRY

ANALYTICAL TRIGONOMETRY

THOMAS J. ROBINSON
ASSOCIATE PROFESSOR OF MATHEMATICS
UNIVERSITY OF NORTH DAKOTA

HARPER & ROW, PUBLISHERS

NEW YORK, EVANSTON, AND LONDON

LIBRARY OF CONGRESS CATALOG CARD NUMBER: 67-10800

CONTENTS

1 Introduction, Review of Algebra   1

v

6 Complex Numbers   124

PREFACE

The primary purpose of this book is to serve as a prerequisite to calculus. For this reason, the amount of computational work is kept at a minimum. To stress the importance of functions, the trigonometric functions are first introduced as being evaluated at real numbers. In fact, angles are not even mentioned until Chapter 4, and then $T(\theta)$, for θ an angle and T a trigonometric function, is defined in terms of $T(t)$ for the same function T and a real number t.

Throughout the book if P is some property possessed by trigonometric functions, the significance of P is investigated in its relation to non-trigonometric functions as well. Such properties as being periodic, or even, or odd, or neither even nor odd, are by no means restricted to the trigonometric functions. As a number of theorems on the above mentioned properties are easy to prove using the definitions of these properties and functional notation, these are also included.

The author has attempted, where possible, to use analytic rather than geometric methods. For this reason the order in which some of the topics appear is different from many trigonometry books. For example, as periodicity occurs within the definitions of the so-called "circular point" and the trigonometric functions, it is discussed at that time. Although periodicity is used later in obtaining the graphs of the trigonometric functions, its use and importance are by no means restricted to graphing. Similarly, the fact that the sine and cosine functions are odd and even, respectively, arises in the development of $\sin (x \pm y)$ and $\cos (x \pm y)$, and these properties are then discussed in general.

As this is primarily a trigonometry book, the first chapter is meant to be only a brief outline of the ideas to be used in the remainder of the book. For this reason relations and mappings are not included. For students who have had a good course in algebra, the first chapter may be omitted. Students beginning in trigonometry or who are taking it concurrently with algebra should find it helpful.

In the chapter on inverse functions and trigonometric equations, Chapter 5, the "converse" of a function is introduced first. While "converse" may not be the best name that could be given, and in fact would not be necessary if relations were used, it is, nevertheless, quite descriptive. A function then has an inverse, which is its converse, if and only if the converse is a function. A one-to-one function is defined, and then it is proved that a function has an inverse, if and only if it is one-to-one. As is true of other concepts in this book, inverses are considered in general at the same time the trigonometric functions are discussed.

In Chapter 4 the standard techniques for solving triangles and finding areas are given.

Complex numbers are first introduced in Chapter 1 as ordered pairs of real numbers with the appropriate definitions of addition and multiplication. Then in Chapter 6, complex numbers are presented in their usual rectangular and polar forms.

Appendix A gives a brief review of logarithms and exponents, and Appendix B, the derivation of the Law of Tangents and tangents of half-angle formulas. Some exercises are included with each appendix. Answers to the odd-numbered problems are given, except that where there are several parts to a problem, only the answers to parts a, c, e, and so on, are included. A teacher's manual is available which includes the answers to the even-numbered problems.

This book is the outgrowth of a set of notes that was used during the years 1964–1966 at the University of North Dakota. Trigonometry is a two semester-hour course there, but the text should suffice for a three semester-hour or three or four quarter-hour course.

The author has, of course, been influenced in this writing by the trigonometry and algebra texts from which he has taught. The integrated algebra and trigonometry book by Fisher and Ziebur and published by Prentice-Hall has probably been the biggest influence.

The author would like to express his appreciation to his colleagues for the suggestions they have made in the preparation of this book. He is especially grateful to Professor Lyle Mauland and Professor Edward Nelson. Professor Mauland and the author taught the first courses from the notes, and he and Professor Nelson provided a great deal of encouragement in the completion of the book.

THOMAS J. ROBINSON

PREFACE TO THE STUDENT

Before he begins the study of this text, the student may benefit from a few comments with regard to what is expected of him. We assume the student is reasonably familiar with the principles of plane geometry including such concepts as arc length and the Pythagorean Theorem. A knowledge of algebra is also desirable, although it should be possible to take this course concurrently with college algebra. An appendix is given that reviews logarithms and exponents.

Throughout the text we have attempted to be precise in the definitions. A number of theorems are included, together with their proofs, in an attempt to broaden the student's mathematical education and, hopefully, help him mature mathematically. As many of these theorems are of the form, "A is true, if and only if B is true" an explanation of this type of statement should be given.

The statement, "A is true, if and only if B is true" contains two statements. They are: "If A is true, then B is true," and "If B is true, then A is true." Thus in proving a theorem of this form, it is necessary to prove both statements. The second is called the converse of the first and vice versa. In the statement, "If A then B," A is called the hypothesis and B is called the conclusion.

The proofs in this text, of both theorems and problems, are generally obtained by one of two methods: constructive, or direct, proof and mathematical induction. A direct proof involves assuming the hypothesis to be true and proving that then the conclusion must also be true, using the hypothesis and other pertinent facts, such as definitions and previously proved theorems.

Mathematical induction is usually used to prove statements involving all positive integers. The principle of mathematical induction is as follows: A statement $S(n)$ is given which involves the positive integer n and the object is to prove that $S(n)$ is true for every positive integer n. First we prove that $S(1)$ is true. Then we assume the statement is true for $n = k$, i.e., $S(k)$ is true, and, from this assumption, prove the statement is true for $n = k + 1$, or $S(k + 1)$ is true. Thus $S(1)$ is true, and if $S(1)$ is true, then $S(2)$ is true; if $S(2)$ is true, then $S(3)$ is true; and so on. The assumption that $S(k)$ is true is called the induction hypothesis.

Another method of proof is obtained by assuming the conclusion is false and proving that then the hypothesis is also false. The statements "If A is true, then B is true" and "If B is false, then A is false" mean exactly the same thing. They are said to be "logically equivalent."

Still another method of proof also arises in mathematics and is called an indirect proof. This type of proof is obtained by assuming the conclusion of the statement to be false and then deriving a result which contradicts some previously established statement. This contradiction then establishes the truth of the conclusion.

ANALYTICAL TRIGONOMETRY

1

INTRODUCTION, REVIEW OF ALGEBRA

1. Real Numbers and Coordinate Lines

Whereas a development of the real numbers is impossible, and indeed undesirable here, we will list some properties enjoyed by them with respect to addition and multiplication. In this section, a, b, and c are arbitrary real numbers, 0 is zero, 1 the integer one, $a + b$ the sum of a and b, and ab the product of a and b. The following is a list of properties:

(A1) $a + b = b + a$ (commutative law).

(A2) $a + (b + c) = (a + b) + c$ (associative law).

(A3) $a + 0 = a$.

(A4)   For each a there is an a' such that $a + a' = 0$.

(M1) $ab = ba$ (commutative law).

(M2) $a(bc) = (ab)c$ (associative law).

(M3) $1a = a$.

(M4)   If $a \neq 0$,   there is an a'' such that $aa'' = 1$.

1

In (A4) the element a' is called the *negative* of a and is written $-a$. In (M4) the element a'' is called the *reciprocal* of a and is written

$$a^{-1} \quad \text{or} \quad \frac{1}{a}.$$

We also call a' and a'' the *additive* and *multiplicative inverse*, respectively, of a. The elements 0 and 1 of (A3) and (M3) are called, respectively, the *additive identity* and *multiplicative identity*.

Another property of the real numbers combines addition and multiplication and is given by

(D1) $a(b + c) = ab + ac$ (distributive law).

Subtraction is defined in terms of addition, and division in terms of multiplication. Thus, $a - b$ is that *unique* number x such that $b + x = a$. Also

$$a \div b = \frac{a}{b}$$

is that *unique* number y such that $by = a$. From this it follows that $a \div 0$ is undefined for $a \neq 0$. As $0y = 0$ for any real number y, when a and b are both zero,

$$a \div b \quad \text{or} \quad \frac{0}{0}$$

is called an *indeterminate* form. Thus in particular we *cannot* say

$$\frac{0}{0} = 0.$$

The nonzero real numbers are divided into two classes—those which are positive and those which are negative. Under the operation of addition, both classes are closed. That is, the sum of two positive numbers is a positive number, and the sum of two negative numbers is a negative number. Under multiplication the positive numbers are closed, but the negative numbers are not, because the product of two negative numbers is a positive number. Of course, the product of a positive number and a negative number is a negative number.

These properties of the real numbers given here are basic in the development of inequalities and other concepts to be considered presently. Before proceeding to inequalities, however, we will give a brief geometric interpretation of the real numbers.

We begin with a horizontal line which we will call L. On L we select an arbitrary point which we label O. This is called the origin and is to correspond to the real number zero. To the right of O select an arbitrary point and label it U. U is called the unit point, corresponds to the integer 1, and the segment from O to U is one unit in length. Using this segment as a standard, we lay off segments of this length to the right of

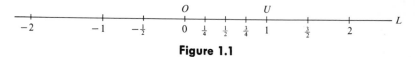

Figure 1.1

U and to the left of O. The end points are labeled 2 and -1, respectively. We continue in this way to lay off segments of length "unity" to the right and left of the points already obtained and label the end points 3, -2, 4, -3, and so on. Using this method we obtain a class of points on the line L, each corresponding to an integer. Each integer is called the *coordinate* of the point to which it corresponds, and we speak of "the point n" when we mean the point with coordinate n.

Next, given any positive integer q, it is possible to divide the segment between O and U into q equal parts. Thus for each integer $p = 1, \cdots$, $q - 1$, we make to correspond a point on the segment which is labeled p/q. Continuing this process between each pair of integral points yields a collection of points labeled m/n for n any positive integer and m any integer. In other words, we have made to correspond to each rational number a point on L. As is the case with the integral points, the other rational numbers are called coordinates of the points to which they correspond, and we equate the points with their coordinates.

Although the above process gives rise to an infinite number of points on the line, it by no means gives a coordinate for each point. In fact there are many more points remaining without coordinates than there are with coordinates. These points are called irrational points and correspond to the irrational numbers. Nevertheless, we will now assume the line L has been so coordinated in some fashion that, given any point P on the line, it has a real number a as its coordinate, and conversely, given any real number b, there is a point Q on the line having b as its coordinate. L is now called a *coordinate line*.

2. Inequalities

Inequalities play a very important role in mathematics. As we wish to use them from time to time in our later work, we will give the necessary definitions and a number of their properties.

Definition 2.1. If a and b are real numbers, a is less than b, written $a < b$, if and only if $b - a$ is a positive number. This may also be written $b > a$ and read "b is greater than a." Using this terminology, we adopt the notation $a > 0$ if a is positive and $a < 0$, if a is negative.

Assuming that every real number is either positive, negative, or zero, but only one of these three, and using the properties of real numbers from section **1**, the following properties are quite easily proved.

Property 2.2. If a and b are any two real numbers, then one and only one of the following holds:

(i) $a < b$.
(ii) $b < a$.
(iii) $a = b$.

Property 2.3. If $a < b$ and c is any real number, then $a + c < b + c$.

Property 2.4. If $a < b$ and $c > 0$, then $ac < bc$.

Property 2.5. If $a < b$ and $c < 0$, then $ac > bc$.

Property 2.6. If $a < b$ and $c < d$, then $a + c < b + d$.

Property 2.7. For any real numbers a, b, and c, if $a < b$ and $b < c$, then $a < c$.

Property 2.2 is called the *trichotomy property* and Property 2.7 is called the *transitive property*. The symbol "$<$" represents an inequality or an "ordering," and the reals are said to be ordered by $<$.

We will indicate the proof of Property 2.5. The others are proved in an analogous manner and are left as exercises.

Proof of Property 2.5. If $a < b$, then by Definition 2.1, $b - a$ is positive, that is, there is a positive number p such that $b - a = p$.

If $c < 0$, then $(b - a)c = bc - ac = pc$, and pc is a negative number. Therefore, $ac - bc = -pc$ is a positive number, and by Definition 2.1, $bc < ac$.

Definition 2.8. If a and b are real numbers, we write $a \leq b$ if and only if a is either less than b or equal to b. As above, $b \geq a$ if and only if b is either greater than a or equal to a.

Properties similar to Properties 2.3–2.7 may now be listed.

Property 2.3'.   If $a \leq b$ and c is any real number, then $a + c \leq b + c$.

Property 2.4'.   If $a \leq b$ and $c > 0$, then $ac \leq bc$.

Property 2.5'.   If $a \leq b$ and $c < 0$, then $ac \geq bc$.

Property 2.6'.   If $a \leq b$ and $c \leq d$, then $a + c \leq b + d$.

Property 2.7'.   For any real numbers a, b, and c, if $a \leq b$ and $b \leq c$, then $a \leq c$.

Another property may be added that combines both "$<$" and "$\leq$" under addition.

Property 2.8'.   If $a < b$ and $c \leq d$, then $a + c < b + d$.

The student should observe that the notation $a \leq b$ does not mean both $a < b$ *and* $a = b$. By Property 2.2 this is impossible. By way of example, however, since $2 < 3$ we may surely write $2 \leq 3$ even though $2 \neq 3$. (The symbol "$a \neq b$" is used to indicate that a is *not* equal to b.) At the same time we write $3 \leq 3$ even though $3 \not< 3$.

Note. The transitive property, Property 2.7, states that if $a < b$ and $b < c$, then $a < c$, and similarly for Property 2.7'. The two inequalities above may be joined to give $a < b < c$, if and only if $a < b$ and $b < c$. Other examples of this type may easily be obtained. Thus $a \leq b < c$, if and only if $a \leq b$ and $b < c$. Both of these examples are called continued inequalities.

The student may have recognized the geometric significance of the inequalities. Using the coordinate line of section 1, it follows that $a < b$, if and only if the point a is to the left of the point b. (Here, once again, the coordinate of the point is used to designate the point itself.) This device is useful in visualizing the solutions of inequalities as explained below.

To solve an inequality involving a number x means to find all the values of x for which the inequality is true. The properties and definitions are used in finding the solution analytically. We will illustrate this with some examples.

Example 2.9. Solve the inequality $x^2 > 1$.

Solution. By Property 2.3, $x^2 > 1$ means the same as $x^2 - 1 > 0$. Factoring this expression as

$$(x - 1)(x + 1) > 0$$

indicates there are separate cases to consider in the completion of the problem, for in section **1** we observed that the product of two numbers is positive, only when both have the same sign. Hence, either $x - 1$ and $x + 1$ are both positive or are both negative.

Case 1. $x - 1 > 0$ and $x + 1 > 0$.
To solve these two inequalities we use Property 2.3 to obtain $x > 1$ and $x > -1$.
But the only way both of these can be satisfied is for $x > 1$.

Case 2. $x - 1 < 0$ and $x + 1 < 0$.
From Property 2.3 we have $x < 1$ and $x < -1$.
But then $x < -1$.
Combining these two cases yields the solution $x > 1$ or $x < -1$.
Geometrically we can proceed as follows. We first consider the factored expression $x^2 - 1 = (x - 1)(x + 1) > 0$ and observe that the individual factors change sign at $+1$ or -1. These points are called critical points. We then divide the coordinate line into three parts—those points to the right of $+1$, those to the left of -1, and those which fall between $+1$ and -1. In each part we indicate the sign on each factor and then use the rules of multiplication to determine the sign of the prod-

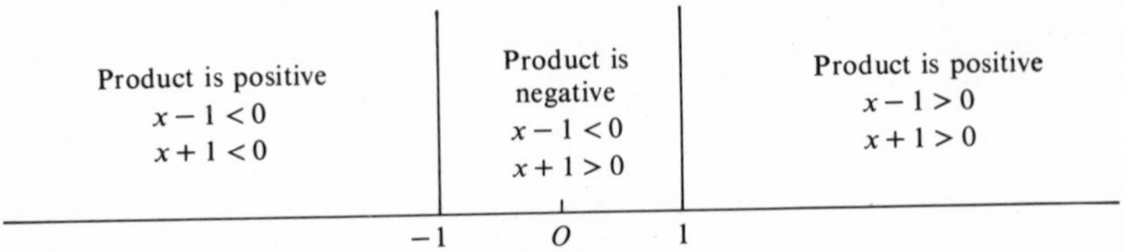

Product is positive $x - 1 < 0$ $x + 1 < 0$	Product is negative $x - 1 < 0$ $x + 1 > 0$	Product is positive $x - 1 > 0$ $x + 1 > 0$
	$-1 \qquad 0 \qquad 1$	

Figure 2.1

uct. Observe, for example, that when x is to the right of $+1$, x is also to the right of -1, so $x - 1$ and $x + 1$ are both positive.

Inequalities involving fractions frequently cause difficulty. The following example can be solved in two different ways.

Example 2.10. Solve

$$\frac{x+1}{x+2} > 3.$$

Solution A. Using Property 2.3, if

$$\frac{x+1}{x+2} > 3,$$

then

$$\frac{x+1}{x+2} - 3 > 0.$$

A common denominator for the left-hand side of the inequality is $x + 2$ and so if

$$\frac{x+1}{x+2} - 3 > 0,$$

then

$$\frac{x + 1 - 3(x+2)}{x+2} = \frac{-2x-5}{x+2} > 0.$$

There are two cases to consider once again, as the quotient of two numbers is positive only if both have the same sign.

Case 1. If $-2x - 5 > 0$ and $x + 2 > 0$, then by Property 2.3, $-5 > 2x$ and $x > -2$. By Property 2.4, $-\frac{5}{2} > x$ and $x > -2$. These two inequalities do not have a common solution, for there is no x such that $x > -2$ and $x < -\frac{5}{2}$.

Case 2. If $-2x - 5 < 0$ and $x + 2 < 0$, then by Property 2.3, $-5 < 2x$ and $x < -2$. By Property 2.4, $-\frac{5}{2} < x$ and $x < -2$, and the solution is $-\frac{5}{2} < x < -2$.

The only solution arises from Case 2.

Solution B. The denominator $x + 2$ of the fraction $(x + 1)/(x + 2)$ is either positive or negative. (Recall that division by zero is excluded.) If $x + 2 > 0$, or $x > -2$, then using Property 2.4 gives:

$$\text{If} \quad (x + 1)/(x + 2) > 3,$$

then

$$x + 1 > 3(x + 2) = 3x + 6.$$

Then by Property 2.3, $-5 > 2x$ or $-\frac{5}{2} > x$. This is the same as Case 1 of Solution A.

Next, if $x + 2 < 0$, or $x < -2$, then using Property 2.5 gives:

$$\text{If } (x + 1)/(x + 2) > 3,$$

then
$$x + 1 < 3(x + 2) = 3x + 6.$$

By Property 2.3, $-5 < 2x$ or $-\frac{5}{2} < x$. Combining this with $x < -2$ gives $-\frac{5}{2} < x < -2$, the same solution as Case 2 of Solution A.

Either of the solutions is correct. It appears that Solution A is a little more satisfactory, as it allows the use of the geometry and critical numbers as indicated in Example 2.9. It also turns out that students who use method B frequently forget that the number $x + 2$ may be negative for some values of x.

One more thing must be pointed out in solving inequalities. The solution is not complete until it is shown that the numbers are actually solutions. Thus, in Example 2.10, it is necessary to show that if $-\frac{5}{2} < x < -2$, then $(x + 1)/(x + 2) > 3$. This, of course, is not difficult, for we write

$$-\tfrac{5}{2} < x < -2$$

implies $-\frac{5}{2} < x$ and $x < -2$.

implies $-5 < 2x$ and $x < -2$.

implies $-5 - 2x < 0$ and $x + 2 < 0$.

implies $(x + 1) - 3(x + 2) < 0$ and $x + 2 < 0$.

implies $x + 1 < 3(x + 2)$ and $x + 2 < 0$.

implies $(x + 1)/(x + 2) > 3$.

Thus it follows that $(x + 1)/(x + 2) > 3$, if and only if $-\frac{5}{2} < x < -2$.

The above procedure is accomplished by using steps in the solution which are reversible, although this is seldom mentioned.

EXERCISES

1. Solve the following inequalities if possible:

(a) $x + 1 > 0$

(b) $x - 2 > 3$

(c) $x + 5 > 2x - 3$

(d) $x + 1 \leq x - 1$

(e) $2x - 5 > x$

(f) $(1/x) > 2$

(g) $x^2 \leq 0$

(h) $x^2 - 4 \leq 0$

(i) $x^2 - 5x + 6 > 0$

(j) $x^2 - 5x + 6 \leq 0$

(k) $x^3 - x \geq 0$

(l) $x^2 + 4x - 2 > x^2 - x + 3$

(m) $x(x + 2)(x - 1) \leq 0$

(n) $(x - 2)(x + 1)(x - 3) > 0$

(o) $x^2 + 2x - 1 > 3$

(p) $(1/x) < 2$

(q) $(x - 1)(x + 2)/(x + 1) < 0$

(r) $(x - 1)/(x - 2) > 2$

(s) $(x + 1)/(2x - 1) < 3$.

2. If $0 < x < 1$, prove $x^3 < x$.

3. If $x > 1$, prove $x^2 > x$.

4. Prove Properties 2.2–2.7 and 2.3′–2.8′.

5. If $0 \leq a < b$, then $a^2 < b^2$.

6. If $0 \leq a$, $0 \leq b$ and $a^2 < b^2$, then $a < b$.

3. Absolute Value

Associated with every real number a is a positive number a' such that if a is positive or zero, $a' = a$, and if a is negative, $a' = -a$. This number a' is called the *absolute value* of a and is written $|a|$. The definition is as follows:

Definition 3.1.

$$|a| = \begin{cases} a & \text{if and only if} \quad a \geq 0; \\ -a & \text{if and only if} \quad a < 0. \end{cases}$$

As examples of this definition we have $|3| = 3$, as $3 > 0$. However, $|-3| = -(-3) = 3$, also.

Note. The student should recognize that $|a| \geq 0$ for every a, and if $a \neq 0$, then $|a| > 0$. An immediate application of the absolute value is obtained by recalling that $\sqrt{a}$ represents the positive number whose square is a, or the positive or principal square root of a. From this and Definition 3.1,

$$\sqrt{x^2} = |x|, \quad not \quad \sqrt{x^2} = \pm x.$$

In particular,

$$\sqrt{(-3)^2} = \sqrt{9} = 3 = |-3|.$$

A common difficulty encountered in the definition of the absolute value is the part involving $|a| = -a$, if and only if $a < 0$. Keep in mind that $-a$ is not always a negative number. In particular, if a is any negative number, then $-a$ must be a positive number.

The following properties of absolute value follow from the definition.

Property 3.2. $|ab| = |a||b|$ for all real numbers a and b.

Property 3.3.

$$\left|\frac{a}{b}\right| = \frac{|a|}{|b|}, \quad b \neq 0.$$

Property 3.4. $|-a| = |a|$.

Property 3.5. $|a|^2 = a^2$.

Inequalities and absolute values are related by the following properties:

Property 3.6. $|x| < b$, if and only if $-b < x < b$.

Property 3.6′. $|x| \leq b$, if and only if $-b \leq x \leq b$.

Property 3.7. $|x| > b$, if and only if $x > b$ or $x < -b$.

Property 3.7′. $|x| \geq b$, if and only if $x \geq b$ or $x \leq -b$.

Property 3.8. $a \leq |a|$ and $-|a| \leq a$.

Using the above properties and the definition of absolute value, it is possible to derive other important inequalities involving absolute values. For example, note the well-known *triangle inequality:*

(3.1) $$|a + b| \leq |a| + |b|,$$

and the inequality

(3.2) $$\big||a| - |b|\big| \leq |a - b|.$$

We will give the proofs of Properties 3.2 and 3.6, and Equation (3.1).

Proof of Property 3.2. By Definition 3.1,

$$|ab| = \begin{cases} ab, & \text{if and only if} \quad ab \geq 0; \\ -(ab), & \text{if and only if} \quad ab < 0. \end{cases}$$

Using the properties of the real numbers under multiplication, $ab \geq 0$ if and only if $a \geq 0$ and $b \geq 0$, or $a \leq 0$ and $b \leq 0$. Then $|ab| = ab = (-a)(-b) = |a||b|$. But also, $ab < 0$, if and only if $a < 0$ and $b > 0$, or $a > 0$ and $b < 0$.
Then

$$-(ab) = |ab| \quad \text{and} \quad -(ab) = (-a)b = |a|\,|b|$$

or $$-(ab) = a(-b) = |a|\,|b|.$$

Thus, $|ab| = |a|\,|b|$, regardless of the signs of a and b.

Proof of Property 3.6.   There are two parts to this proof. First it is necessary to prove that, if $|x| < b$, then $-b < x < b$, and second, if $-b < x < b$, then $|x| < b$.
Suppose first that $|x| < b$. By definition,

$$|x| = \begin{cases} x, & \text{if and only if } x \geq 0; \\ -x, & \text{if and only if } x < 0. \end{cases}$$

Hence, if $|x| < b$, then $0 \leq x < b$, when $x \geq 0$ and clearly $-b < 0 \leq x < b$. Now if $|x| < b$, then $0 < -x < b$ when $x < 0$, and hence, $-b < x < 0 < b$. In either case the conclusion follows.
Suppose next that $-b < x < b$. If $0 \leq x < b$, then $x = |x| < b$. If $-b < x < 0$, then $0 < -x = |x| < b$.

Proof of Equation (3.1).   By Property 3.8, if a and b are any real numbers, then

$$-|a| \leq a \leq |a|$$

and $$-|b| \leq b \leq |b|.$$

Then by Property 2.6′, $-(|a| + |b|) \leq a + b \leq |a| + |b|$. By Property 3.6′, $|a + b| \leq ||a| + |b||$. But $|a| + |b| \geq 0$ so $|a + b| \leq |a| + |b|$.
In the previous section we considered the solution of inequalities. In the same manner, the solution of an inequality or an equation in $|x|$ consists of the real numbers satisfying that inequality or equation. The following examples illustrate some methods that may be used.

Example 3.9.   Solve $|2x - 1| = 3$.

Solution. By Definition 3.1, $|2x - 1| = 3$, if and only if $2x - 1 = 3$ or $2x - 1 = -3$. Then if $2x - 1 = 3$, $2x = 4$, or $x = 2$. If $2x - 1 = -3$, $2x = -2$ or $x = -1$. Then the solutions are $x = 2, -1$. It should be noted we have only shown that if $|2x - 1| = 3$, then $x = 2$ or -1. To complete the solution, observe that if $x = 2$, then $|2x - 1| = |2 \cdot 2 - 1| = |3| = 3$, and if $x = -1$, then $|2x - 1| = |2(-1) - 1| = |-3| = 3$.

Example 3.10. Solve $|3x + 2| \leq 4$.

Solution. By Property 3.6′, $|3x + 2| \leq 4$, if and only if $-4 \leq 3x + 2 \leq 4$. Then using Property 2.3′, $-6 \leq 3x \leq 2$, and by Property 2.4′, $-2 \leq x \leq \frac{2}{3}$.

Example 3.11. Solve $|x + 1| \leq |x - 2|$.

There are two methods that can be used to solve this inequality, one using Definition 3.1 and the other using Property 3.5. The second of these two is the more convenient one.

Solution A. Using Definition 3.1,

$$|x + 1| = \begin{cases} x + 1, & \text{if and only if } x + 1 \geq 0; \\ -(x + 1), & \text{if and only if } x + 1 < 0; \end{cases}$$

and

$$|x - 2| = \begin{cases} x - 2 & \text{if and only if } x - 2 \geq 0; \\ -(x - 2) & \text{if and only if } x - 2 < 0. \end{cases}$$

Thus to find the numbers x such that $|x + 1| \leq |x - 2|$, we consider the different possibilities arising from the above relations. There are four cases to consider:

Case 1. $x + 1 \leq x - 2$, $x + 1 \geq 0$ and $x - 2 \geq 0$. Clearly $x + 1 \leq x - 2$ for no x satisfying $x + 1 \geq 0$ and $x - 2 \geq 0$.

Case 2. $-(x + 1) \leq -(x - 2)$, $x + 1 < 0$ and $x - 2 < 0$. Then $-x - 1 \leq -x + 2$, which is true for all x satisfying $x + 1 < 0$ and $x - 2 < 0$. Hence, if $x < -1$, then $|x + 1| \leq |x - 2|$.

Case 3. $x + 1 \leq -(x - 2)$, $x + 1 \geq 0$ and $x - 2 < 0$. Then $x + 1 \leq -x + 2$ or $2x \leq 1$ and $x \leq \frac{1}{2}$. Combining this with the above inequalities gives $-1 \leq x \leq \frac{1}{2}$.

Case 4. $-(x + 1) \leq x - 2$, $x + 1 < 0$ and $x - 2 \geq 0$. There is no solution here, for then $x < -1$ and $x \geq 2$.

From Cases 2 and 3 it follows that the solution of the inequality is $x \leq \frac{1}{2}$.

Solution B.   Using Problem 5 of section 2, it follows that if $|x + 1| \leq |x - 2|$, then $|x + 1|^2 \leq |x - 2|^2$, as $|a| \geq 0$ for every real number a. By Property 3.5, $|x + 1|^2 = (x + 1)^2 = x^2 + 2x + 1$ and $|x - 2|^2 = (x - 2)^2 = x^2 - 4x + 4$. Thus $x^2 + 2x + 1 \leq x^2 - 4x + 4$, or $6x \leq 3$, and finally $x \leq \frac{1}{2}$. Using Problem 6 of section 2, all of the operations can be reversed, and we may write $|x + 1| \leq |x - 2|$, if and only if $x \leq \frac{1}{2}$.

The absolute value can be interpreted geometrically as the distance between points on a coordinate line. Thus, $|a - b|$ represents the distance between the points whose coordinates are a and b, respectively. In particular $|x - 1|$ is the distance from a point whose coordinate is x to the unit point; $|x|$ represents the distance from the point with coordinate x to the origin.

EXERCISES

1. Solve the following:
   (a) $|x + 2| = 3$ (e) $|2x - 1| < 3$ (i) $|2d - x| > 3$
   (b) $|x| + 1 = 2$ (f) $|3x + 2| \leq 4$ (j) $|x + 1| \leq |x| + 1$
   (c) $|x^2 - 2x| = 1$ (g) $|x + 1| > 2$ (k) $|x + 1| \leq |x|$.
   (d) $|x| < 1$ (h) $|3x - 4| > 1$

2. Prove Properties 3.3–3.5, 3.6′, 3.7, 3.7′, and 3.8.

3. Prove $\big||a| - |b|\big| \leq |a - b|$. (*Hint.* $|a| = |a - b + b|$. Then use the triangle inequality.)

4. Rectangular Coordinates and the Distance Formula

Let L_1 and L_2 be two coordinate lines that lie in a plane and intersect at right angles at their origins. For simplicity we assume that their unit segments are identical, that L_1 is horizontal with its positive part to the right of L_2, and the positive part of L_2 is above L_1.

Let P be any point in the plane. From the point P drop perpendiculars to the lines L_1 and L_2. The intersection point (coordinate) on L_1 is indicated by x and the intersection point (coordinate) on L_2 by y. The number x is called the *abscissa* and the number y the *ordinate* of P, and together they are called the *coordinates* of P. The coordinates are listed together as (x,y) (later to be called an ordered pair of real numbers).

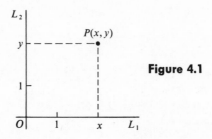

Figure 4.1

Given any point P in the plane, there is exactly one pair of real numbers (x,y) corresponding to P and conversely, given any pair (x,y) of real numbers, there is exactly one point in the plane whose abscissa is x and whose ordinate is y.

As a point on a coordinate line is identified with its coordinate, so a point in the plane is identified with its coordinates. Thus one says "the point $P(x,y)$" instead of "the point P with coordinates (x,y)." The coordinate lines L_1 and L_2 are called the *coordinate axes* and are usually labeled X (or x) and Y (or y), respectively. The plane with the coordinate axes is called the *coordinate plane* or the *Cartesian plane*.

The x and y axes divide the plane into four parts, called *quadrants*. If $P(x,y)$ is any point in the plane and not on a coordinate axis, then the coordinates of P have the signs as indicated in the figure below.

If P_1 and P_2 are points in the plane, the distance between them is indicated by the symbol $|P_1P_2|$. This is not to be confused with the symbol

	Y	
II		I
$x < 0$		$x > 0$
$y > 0$		$y > 0$
	O	X
III		IV
$x < 0$		$x > 0$
$y < 0$		$y < 0$

Figure 4.2

"| |" when used to represent absolute value. The following theorem gives a formula for finding the distance between points in the plane:

Theorem 4.1. If $P_1(x_1,y_1)$ and $P_2(x_2,y_2)$ are any two points in the plane, then the distance between P_1 and P_2 is given by

(4.1) $$|P_1P_2| = \sqrt{(x_2 - x_1)^2 + (y_2 - y_1)^2}.$$

Proof. The position of the points is immaterial, so for simplicity let P_1 be to the left of P_2. (See Figure 4.3.) A horizontal line through P_1 and a vertical line through P_2 will intersect at right angles at the point Q whose coordinates are (x_2,y_1). Then by the Pythagorean theorem, it follows that

$$|P_1P_2|^2 = |P_1Q|^2 + |P_2Q|^2.$$

It should be noted that

$$|P_1Q| = |P_1'P_2'| = |x_2 - x_1|,$$

and $$|P_2Q| = |P_1''P_2''| = |y_2 - y_1|.$$

Thus,

$$|P_1P_2|^2 = |x_2 - x_1|^2 + |y_2 - y_1|^2 = (x_2 - x_1)^2 + (y_2 - y_1)^2$$

by Property 3.5. Extracting the square root from each side yields

$$|P_1P_2| = \sqrt{(x_2 - x_1)^2 + (y_2 - y_1)^2}$$

and the theorem is proved.

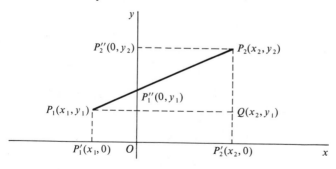

Figure 4.3

Using the distance formula as applied to $P_1(2,-1)$ and $P_2(5,4)$, it follows that

$$|P_1P_2| = \sqrt{(x_2 - x_1)^2 + (y_2 - y_1)^2}$$
$$= \sqrt{(5-2)^2 + [4 - (-1)]^2}$$
$$= \sqrt{3^2 + 5^2}$$
$$= \sqrt{34}.$$

The following is an example of how the distance formula may be applied:

Example 4.2. The points $A(2,3)$, $B(6,10)$, and $C(10,4)$ are the vertices of an isosceles triangle.

$$|AB| = \sqrt{(6-2)^2 + (10-3)^2} = \sqrt{16 + 49} = \sqrt{65},$$

and $\quad |AC| = \sqrt{(10-2)^2 + (4-3)^2} = \sqrt{64 + 1} = \sqrt{65}.$

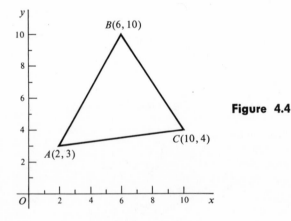

Figure 4.4

Three points are said to be *collinear* if they lie in the same straight line. Using the distance formula, three points P_1, P_2, and P_3, with P_2 between P_1 and P_3, are collinear if $|P_1P_2| + |P_2P_3| = |P_1P_3|$.

Example 4.3. Determine whether the points $P_1(1,3)$, $P_2(3,7)$, and $P_3(4,9)$ are collinear.

Solution.

$$|P_1P_2| = \sqrt{(3-1)^2 + (7-3)^2} = \sqrt{4+16} = \sqrt{20} = 2\sqrt{5},$$

$$|P_2P_3| = \sqrt{(4-3)^2 + (9-7)^2} = \sqrt{1+4} = \sqrt{5},$$

$$|P_1P_3| = \sqrt{(4-1)^2 + (9-3)^2} = \sqrt{9+36} = \sqrt{45} = 3\sqrt{5}.$$

Thus

$$|P_1P_2| + |P_2P_3| = 2\sqrt{5} + \sqrt{5} = 3\sqrt{5} = |P_1P_3|,$$

and the points are collinear.

EXERCISES

1. Find the distance between the following pairs of points:
   (a) $(1,0)$, $(3,8)$ (c) $(1,27)$, $(-4,15)$ (e) $(2,1)$, $(3,7)$
   (b) $(3,4)$, $(2,-5)$ (d) $(-5,4)$, $(1,2)$ (f) $(3,-1)$, $(-2,-2)$.
2. Determine whether the following points are vertices of an isosceles triangle:
   (a) $A(1,3)$, $B(5,4)$, $C(7,4)$
   (b) $A(2,4)$, $B(5,1)$, $C(-1,7)$
   (c) $A(5,3)$, $B(6,11)$, $C(9,10)$.
3. Determine whether the following points are vertices of a right triangle:
   (a) $A(2,-1)$, $B(6,6)$, $C(6,-1)$
   (b) $A(1,1)$, $B(5,-3)$, $C(4,4)$
   (c) $A(-2,-1)$, $B(4,3)$, $C(3,-3)$.
4. Determine whether the points are collinear:
   (a) $(-1,-3)$, $(0,0)$, $(2,6)$
   (b) $(-2,4)$, $(3,6)$, $(5,7)$
   (c) $(3,1)$, $(6,4)$, $(8,6)$
   (d) $(0,2)$, $(3,-4)$, $(4,-6)$.
5. If the opposite sides of a quadrilateral are equal, the quadrilateral is a parallelogram. If the adjacent sides of a parallelogram are equal, it is a rhombus, and if the diagonals of a parallelogram are equal, it is a rectangle. Determine whether the following points are the vertices of a parallelogram, rhombus, rectangle, square, or none of these.
   (a) $(1,0)$, $(4,0)$, $(5,5)$, $(2,5)$
   (b) $(-1,1)$, $(3,2)$, $(0,-3)$, $(4,-2)$
   (c) $(0,0)$, $(3,4)$, $(5,8)$, $(2,4)$
   (d) $(0,-1)$, $(3,0)$, $(1,2)$, $(-2,2)$.

5. Sets

Functions will be introduced in terms of sets, and so we will give the background material needed. The term "set" is an undefined concept, but it will be taken as synonymous with aggregate or collection of elements. If A is a set and a is an element of A, we will write $a \in A$.

There are two common ways of representing sets. One is to list all the members or elements. For example, the set $S = \{1, 2, 3, 4\}$, or S is the set consisting of the integers 1,2,3, and 4. We note in passing that changing the order in which the elements are written does not alter the set. Thus the set S above is the same as the set $\{1, 3, 4, 2\}$.

Listing the elements to describe a set works quite well when considering sets with very few elements. However, even for large finite sets, this is not very satisfactory, and for infinite sets it is impossible. Hence, it is frequently more satisfactory to represent a set by indicating a relation among its elements. For example, $S = \{x : x$ is a positive integer, $x \leq 4\}$, read "S is the set of all x such that x is a positive integer and x is less than or equal to 4." In this case, of course, there is no advantage in building up a set by this second method, but the set I_{1000} of the first 1000 positive integers can easily be given by $I_{1000} = \{x : x$ is a positive integer, $x \leq 1000\}$.

For infinite sets it is not possible to list all the elements, and hence, the second method of describing sets is used. For example, the set Q of rational numbers can be given by $Q = \{m/n : m,n$ are integers, $n > 0\}$. The integers, the positive integers, and the set of all real numbers are sets which are encountered frequently. They will be represented by I, I^+, and R, respectively.

Some common sets of real numbers and the special symbols representing them are the following:

(i) The "closed interval" $[a,b] = \{x : a \leq x \leq b\}$.

(ii) The "open interval" $(a,b) = \{x : a < x < b\}$.

(iii) The "closed-open interval" $[a,b) = \{x : a \leq x < b\}$.

(iv) The "open-closed interval" $(a,b] = \{x : a < x \leq b\}$.

(v) The "left open ray" $(-\infty,a) = \{x : x < a\}$.

(vi) The "left closed ray" $(-\infty,a] = \{x : x \leq a\}$.

(vii) The "right open ray" $(b,\infty) = \{x : x > b\}$.

(viii) The "right closed ray" $[b,\infty) = \{x : x \geq b\}$.

Before continuing, let us emphasize again the symbolism used to describe a set. We write $\{x : x$ has property $P\}$. "$\{x$" is read "the set of all x," the colon is read "such that," and the statement "x has property P" describes the relation which holds among these members x. A vertical bar is sometimes used instead of the colon to indicate "such that."

The notion of "subset" is an important one, and will be given before proceeding to some examples of sets in the plane.

Definition 5.1. Let A and B be sets. A is a *subset* of B, written $A \subset B$, if and only if every element of A is an element of B. Symbolically, $A \subset B$, if and only if for every $x \in A$ it is also true that $x \in B$. This, of course, does not exclude the possibility that $A = B$. It is true, however, that $A = B$, if and only if $A \subset B$ and $B \subset A$.

Some examples of sets in the coordinate plane, or in view of Definition 5.1, subsets of the coordinate plane, are given below. (x,y) will also indicate the point having coordinates (x,y). The geometric realization of such a set is the graph of the set.

Example 5.2. The circle with radius 1 and with center at the origin, the so-called unit circle, is $C = \{(x,y) : x^2 + y^2 = 1\}$.

Example 5.3. The unit disk with center at the origin is $D = \{(x,y) : x^2 + y^2 \leq 1\}$.

Example 5.4. Find the graph of the set $\{(x,y) : |x| \leq 1\}$.

Solution. This is the set of all points (x,y) in the plane with $|x| \leq 1$, or $-1 \leq x \leq 1$, and y unrestricted. The graph of this set is an infinite strip, the shaded part in the figure below.

Figure 5.1

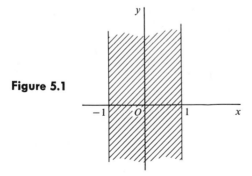

Example 5.5. Find the graph of the set $\{(x,y) : |x| \leq 1 \text{ and } |y| \leq 1\}$.

Solution. This is the set of all points (x,y) in the plane with both $-1 \leq x \leq 1$ and $-1 \leq y \leq 1$. The graph is the square shaded region in the figure below.

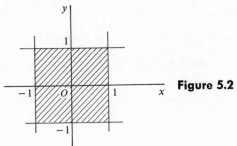

Figure 5.2

EXERCISES

1. Find the following sets of real numbers on a coordinate line:
   (a) $\{-1, -2, 1, 2, 3, 5\}$
   (b) $\{x : x = 1/n, n \in I^+\}$
   (c) $\{x : x \geq 0 \text{ and } x \leq 3\}$
   (d) $\{x : x^2 - 3x - 4 = 0\}$
   (e) $\{x : x > 3 \text{ and } x < 7\}$
   (f) $\{x : (x - 1)(x + 2) \leq 0\}$
   (g) $\{x : (x + 2)(2x - 1) \geq 0\}$
   (h) $\{x : -1 \leq x < 2 \text{ or } 3 \leq x < 4\}$
   (i) $\{x : 0 \leq x < 2 \text{ and } 1 \leq x < 3\}$.

2. Find the graph of the set C of Example 5.2.

3. Find the graph of the set D of Example 5.3.

4. Find the graph of each of the following sets of points in the plane:
   (a) $\{(x,y) : y = 2x + 1, x \in \{-1, 1, 2, 3\}\}$
   (b) $\{(x,y) : y = x + 4, x \in \{-3, -2, -1, 0, 1, 2\}\}$
   (c) $\{(x,y) : y = 3 - x, x \in \{-2, -1, 0, 1, 2, 3\}\}$
   (d) $\{(x,y) : y = x, 0 \leq x < 3\}$
   (e) $\{(x,y) : y = |x|, -3 \leq x \leq 4\}$
   (f) $\{(x,y) : 1 < x \leq 3, 2 \leq y < 4\}$
   (g) $\{(x,y) : y = 3x, -2 \leq x \leq 1\}$
   (h) $\{(x,y) : |x| + |y| = 1\}$
   (i) $\{(x,y) : |x| + |y| \leq 1\}$.

6. Functions

Let A and B be sets. The symbol (a,b), where $a \in A$ and $b \in B$ is called an ordered pair. Equality of ordered pairs is defined by the statement $(a,b) = (c,d)$, if and only if $a = c$ and $b = d$.

Definition 6.1. Let A and B be sets and $F \subset \{(a,b) : a \in A, b \in B\}$. Then F is called a *function*, if and only if for each x: if $(x,y) \in F$ and $(x,z) \in F$, then $y = z$. If F is a function and $(x,y) \in F$, then we write $y = F(x)$ and say "y is the value of the function F at x."

Note. There are two alternate ways of formulating the definition of a function. One is to say it is a set of ordered pairs, no two of which have the same first elements. A second would be to say that F is *not* a function, if and only if for some $a \in A$ and $b, c \in B$ with $b \neq c$, the pairs (a,b) and (a,c) are both in F.

Definition 6.2. Let F be a function. The *domain* and *range* of F are defined by domain $F = \{x : \text{for some } y, (x,y) \in F\}$, and range $F = \{y : \text{for some } x, (x,y) \in F\}$.

We now consider several examples of sets of ordered pairs and apply Definition 6.1 to determine whether or not they are functions.

Example 6.3. $g = \{(1,2), (2,3), (2,4), (3,2)\}$. g is not a function since $(2,3) \in g$ and $(2,4) \in g$, but $3 \neq 4$.

Example 6.4. $f = \{(1,4), (3,2), (5,9)\}$. f is clearly a function, for there are only three ordered pairs in f, and the first elements are all different. Here the domain of f is the set $\{1,3,5\}$ and the range of f is $\{2,4,9\}$. Observe that $f(1) = 4$, $f(3) = 2$, and $f(5) = 9$.

Example 6.5. $h = \{(7,2), (8,2), (9,3)\}$. h is a function, for again there are only three ordered pairs, the first elements of which are distinct. Domain $h = \{7,8,9\}$ while range $h = \{2,3\}$.

In Example 6.5 the integer "2" is the second element of two ordered pairs. There is no restriction in Definition 6.1 of the second elements of ordered pairs; only the first elements cannot be repeated.

Some slightly more complicated examples of sets of ordered pairs follow below.

Example 6.6. $S = \{(x,y) : x \in R, y = x^2\}$. S is a function, for if $(x,y), (x,z) \in S$, then $y = x^2$ and $z = x^2$ which imply $y = z$. Domain S is the set of all real numbers, and range S is the set of all non-negative real numbers. S is called the "square" function. Observe that for each $(x,y) \in S$, $y = S(x) = x^2$. Hence, we could write $S = \{(x,x^2) : x \text{ is a real number}\}$. Some particular examples of values of S are $S(1) = 1^2 = 1$; $S(-2) = (-2)^2 = 4$; $S(a) = a^2$.

Example 6.7. $g = \{(x,y) : x \geq 1, \ y = \sqrt{x-1}\}$. g is a function, for if (x,y), $(x,z) \in g$, then $y = \sqrt{x-1}$ and $z = \sqrt{x-1}$, which imply $y = z$. Domain $g = [1,\infty)$ and range $g = [0,\infty)$. For each $(x,y) \in g$, $y = g(x) = \sqrt{x-1}$. Hence, g may be written as $g = \{(x, \sqrt{x-1}) : x \geq 1\}$. Some examples of values are $g(1) = \sqrt{1-1} = \sqrt{0} = 0$; $g(2) = \sqrt{2-1} = 1$; $g(5) = \sqrt{5-1} = \sqrt{4} = 2$; $g(4) = \sqrt{4-1} = \sqrt{3}$.

In Example 6.7, if the range of g is to consist only of real numbers, the domain of g is the set of all real numbers for which the equation relating the first and second elements of ordered pairs is then defined. For many functions the domain is found in precisely this manner. Hence, if the domain is not specified, simply let the domain be that set of numbers for which the relation is defined. Thus g of Example 6.7 may be written as $g = \{(x,y) : y = \sqrt{x-1}, y \in R\}$. Incidentally, the domain is often called "domain of definition" of the function, and this discussion illustrates that terminology.

Example 6.8. $h = \{(x,y) : y^2 = x, \ x \geq 0\}$. In particular suppose $x > 0$ and (x,y), $(x,z) \in h$. Then $y^2 = x$ and $z^2 = x$. y and z are both nonzero. The two equations above lead to $y^2 = z^2$, or $y^2 - z^2 = 0$. Then $(y+z)(y-z) = 0$. Hence, $y + z = 0$ or $y - z = 0$, i.e., $z = y$ or $z = -y$, and it should be clear that, without making some restriction on the second elements of the pairs in h, there are two possible second elements for each nonzero first element. In particular, the pairs $(1,1)$ and $(1,-1)$ are both in h.

By restricting the values of y sufficiently, one can construct a set which looks like h but is a function.

Example 6.9. $h' = \{(x,y) : y^2 = x, \ x \geq 0, \ y \geq 0\}$. If $(x,y) \in h'$ and $(x,z) \in h'$, then $y^2 = x$ and $z^2 = x$, with both $y,z \geq 0$. Then $y^2 = z^2$, or $y^2 - z^2 = 0$. But now $(y-z)(y+z) = 0$, $y,z \geq 0$. Hence, $y - z = 0$ or $y + z = 0$, i.e., $y = z$ or $y = -z$. But if y and $z \geq 0$, $y = -z$ is impossible for $y \neq 0$. Hence, $y = z$ and h' is a function.

Example 6.10. $\sqrt[3]{\ } = \{(x,y) : y^3 = x\}$. Let $(x,y) \in \sqrt[3]{\ }$ and $(x,z) \in \sqrt[3]{\ }$. Then $y^3 = x$ and $z^3 = x$. It follows that $y^3 = z^3$, or $y^3 - z^3 = 0$. Then $(y-z)(y^2 + zy + x^2) = 0$. Hence, $y - z = 0$ or $y^2 + zy + z^2 = 0$. However, there are no nonzero real numbers y and z such that $y^2 + zy + z^2 = 0$, and so $y = z$. Thus $\sqrt[3]{\ }$ is a function, called the "cube root" function.

Sums, differences, products, and quotients of functions are often encountered. The sum and product are defined below.

Definition 6.11. Let f and g be functions. Then $f + g$ and $f \cdot g$ are defined by

(a) $f + g = \{(x, a + b) : (x, a) \in f$ and $(x, b) \in g\}$,
(b) $f \cdot g = \{(x, a \cdot b) : (x, a) \in f$ and $(x, b) \in g\}$.

One should observe that domain $(f + g)$ and domain $(f \cdot g)$ are the same and consist only of those x which are in both domain f and domain g. From the definition one writes $(f + g)(x) = f(x) + g(x)$ and $(f \cdot g)(x) = f(x)g(x)$, or the value of the sum is the sum of the values, and the value of the product is the product of the values. If $f = \{(x, y) : y = x^2\}$ and $g = \{(x, y) : y = 2x\}$, then $f + g = \{(x, y) : y = x^2 + 2x\}$ and $f \cdot g = \{(x, y) : y = 2x^3\}$.

In Examples 6.6, 6.7, 6.9, 6.10, and the f and g of the previous paragraph, the functions are defined by an equation relating the first and second elements of the ordered pairs. This leads to a special class of functions defined as follows.

Definition 6.12. If $y = f(x)$ is an equation which has been solved for y in terms of x, and $f = \{(x, y) : y = f(x)\}$ is a function, then f is called the *function defined by the equation* $y = f(x)$.

Most of the functions considered in this book are of the type indicated by Definition 6.12. For example, the function f defined by the equation $f(x) = 2x^2 - x + 1$ is $f = \{(x, y) : y = f(x) = 2x^2 - x + 1\}$.

To conclude this section we give an interesting operation involving functions. We first illustrate the operation and then formally define it.

Example 6.13. Let S be defined by $S(x) = x^2$ and f by

$$f(x) = \frac{x + 1}{x - 1}, \quad x \neq 1.$$

The equations $S[f(x)] = [f(x)]^2 = \left(\frac{x + 1}{x - 1}\right)^2, \quad x \neq 1,$

and $f[S(x)] = \frac{S(x) + 1}{S(x) - 1} = \frac{x^2 + 1}{x^2 - 1}, \quad x^2 \neq 1$

are obtained by direct substitution. In the first case $f(x)$ is substituted for x in the equation for $S(x)$, and in the second, $S(x)$ is substituted for x in the equation for $f(x)$. Some numerical examples are $S(2) = 4$, so

$$f[S(2)] = f(4) = \frac{4+1}{4-1} = \frac{2^2+1}{2^2-1} = \tfrac{5}{3},$$

$$f(2) = \frac{2+1}{2-1} = 3,$$

and

$$S[f(2)] = S(3) = \left(\frac{2+1}{2-1}\right)^2 = 9.$$

Definition 6.14. Let f and g be functions. Then the function $f \circ g$, read f composed with g, is defined by $f \circ g = \{(x,z) : \text{for some } y, (x,y) \in g \text{ and } (y,z) \in f\}$. Thus, if $h = f \circ g$, then $h(x) = f[g(x)]$. It should be noted that there must be elements in range g which are also in domain f. The domain of $f \circ g$ is $\{x : x \in \text{domain } g \text{ and } g(x) \in \text{domain } f\}$.

An error that is often made is to confuse the function with its values. $f(x)$ means only the value of the function f at x and is just one element in range f.

EXERCISES

1. Determine which of the following sets of ordered pairs of real numbers represent functions. If the set is a function, specify its domain and range.
   (a) $f = \{(1,3), (32,4), (2,3), (4,2), (2,5)\}$
   (b) $g = \{(3,6), (2,6), (6,6), (5,6)\}$
   (c) $h = \{(x,y) : y = x^3, x \in R\}$
   (d) $I = \{(x,y) : y = x, x \in R\}$
   (e) $F = \{(x,y) : y^3 + x^3 = 1\}$
   (f) $G = \{(x,y) : x^2 + y^2 = 1\}$.

2. If f is the function defined by the equation $f(x) = 2x^2 - 3x + 5$, find $f(0)$, $f(1)$, $f(-1)$, $f(-2)$, $f(a)$, $f(a+1)$, $f(a+h)$.

3. If g is defined by $g(x) = (x - 3)/(x - 2)$, find $g(0)$, $g(3)$, $g(-2)$, $g(1)$, $g(4)$. What is domain g?

4. If f is the function of Problem 2 and g the function of Problem 3, find the following: $f[g(1)]$, $f[g(-x)]$, $g[f(-1)]$, $g[f(x)]$.

5. Write a definition for the division of functions.

6. Write a definition for the sum of the functions f, g, and h.

7. If f and g are real functions, prove $f + g = g + f$.

7. Graphs of Functions and Some Special Functions

Definition 7.1.   The *graph* of an equation relating x and y is the set of all points $P(x,y)$, whose coordinates satisfy the equation.

Definition 7.2.   Let f be a function whose domain and range are sets of real numbers. Then the *graph* of f in the plane is the set of all points $P(x,y)$, where $y = f(x)$.

Thus, if f is the function defined by the equation $y = f(x)$, then the graph of f and the graph of this equation must coincide. Using Definition 7.2 and the definition of a function, it follows that no vertical line crosses the graph of a function in more than one point, assuming the domain to be on the x axis.

Example 7.3.   Let S be defined by $S(x) = x^2$. Its graph is given below, together with a few points on the graph.

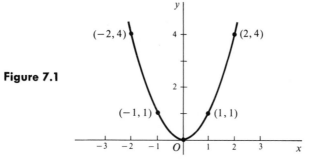

Figure 7.1

Example 7.4.   The *greatest integer function* is defined by $[x]$ = the greatest integer $\le x$. For $0 \le x < 1$, $[x] = 0$, for $1 \le x < 2$, $[x] = 1$; for $2 \le x < 3$, $[x] = 2$; for $-1 \le x < 0$, $[x] = -1$, etc. Its graph is the following:

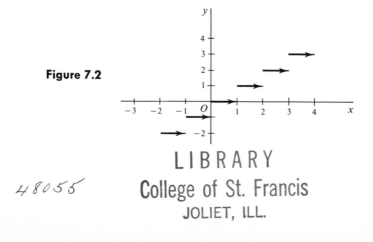

Figure 7.2

Example 7.5.   The *absolute value function* is defined by the equation $f(x) = |x|$. Its graph follows.

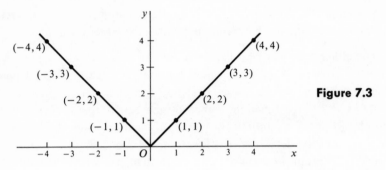

Figure 7.3

Example 7.6.   The *identity function* is defined by $I(x) = x$.

Figure 7.4

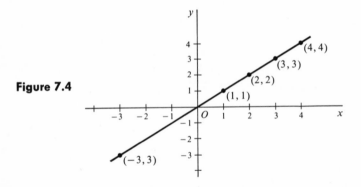

Example 7.7.   The *exponential function* to the base a, where $a > 0$ and $a \neq 1$, is defined by the equation $\exp_a(x) = a^x$. This is well-defined for any rational number x, and it can be extended to all real numbers in such a way that all the rules of exponents are valid. We consider the graphs of two of these functions on the same coordinate system to illustrate the difference when $0 < a < 1$ or $a > 1$. See Appendix A.

Example 7.8.   The *logarithm* of x to the base a, written $\log_a x$, is defined to be the exponent to which a is raised in order to get x. Thus $y = \log_a x$ means the same as $x = a^y$. Then the *logarithm function* to the base a is defined by the equation $L(x) = \log_a x$. As in the exponential function, $a > 0$ and $a \neq 1$. See Appendix A.

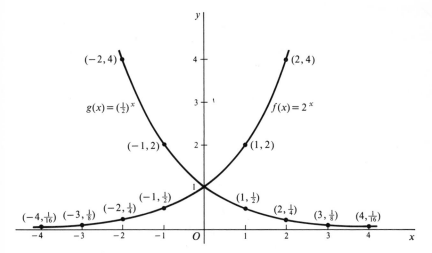

Figure 7.5

Note that the domain of the logarithm function is the same as the range of the exponential function, and conversely, the range of the logarithm function is the same as the domain of the exponential function. A sketch of the graph of the log function with base 2 is given below.

Figure 7.6

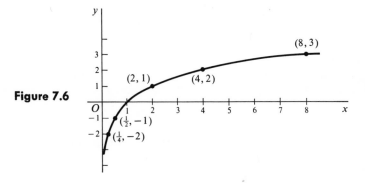

EXERCISES

1. Sketch the graphs of the following functions.
    (a) $f = \{(1,2), (3,4), (4,6)\}$
    (b) $f = \{(2,4), (3,4), (4,4)\}$
    (c) $f = \{(x,y) : y = 3x + 2, x \in \{-2,-1,0,1,2,3\}\}$
    (d) $f = \{(x,y) : y = x - 3, x \in \{-2,-1,1,3,4\}\}$

(e) $f = \{(x,y) : y = x^2, x \in \{1,2,3,4\}\}$
(f) $f = \{(x,y) : y = 4x - 1\}$
(g) $f = \{(x,y) : y = -x\}$
(h) $f = \{(x,y) : y = -x^2\}$
(i) $f = \{(x,y) : y = 3^x\}$
(j) $f = \{(x,y) : y = 2^{x^2}\}$
(k) $f = \{(x,y) : y = \log_3 x\}$.

2. Use Definition 6.11(a), the definition of the sum of functions, to sketch the graphs of the following:

(a) $f = \{(x,y) : y = x^2 + 1, x \in \{-2,-1,0,1,2,3\}\}$. *Hint:* Let $g = \{(x,y) : y = x^2, x \in \{-2,-1,0,1,2,3\}\}$ and $h = \{(x,y) : y = 1, x \in \{-2,-1,0,1,2,3\}\}$. Then $f = g + h$, or $f(x) = g(x) + h(x)$. Sketch g, h, and their sum.

(b) $f = \{(x,y) : y = x + \log_2 x, x \in \{\frac{1}{4},\frac{1}{2},1,2,4\}\}$
(c) $f = \{(x,y) : y = 2x^2 - 3x, x \in \{-1,0,1,2,3,4\}\}$
(d) $f = \{(x,y) : y = x^2 - 1\}$
(e) $f = \{(x,y) : y = x^2 + 2x\}$
(f) $f = \{(x,y) : y = x^3 - x\}$.

3. Let $f = \{(1,2), (3,4), (4,3)\}$ and $g = \{(2,2), (3,2), (4,1)\}$. Sketch the graphs of $f \circ g$ and $g \circ f$, first sketching f and g.

4. Show that if I is the identity function, the function of Example 7.6, and f is any function, then $f \circ I = I \circ f = f$.

8. Complex Numbers

A brief introduction to complex numbers in terms of ordered pairs of real numbers is given below. Complex numbers will be taken up again at greater depth in Chapter 6.

Definition 8.1. A *complex number* is an ordered pair (a,b) of real numbers. Two complex numbers (a,b) and (c,d) are *equal*, $(a,b) = (c,d)$, if and only if $a = c$ and $b = d$.

Definition 8.2. The sum and product of two complex numbers are defined by the respective equations

$$(8.1) \qquad (a,b) + (c,d) = (a + c, b + d),$$

$$(8.2) \qquad (a,b) \cdot (c,d) = (ac - bd, ad + bc).$$

Some numerical examples of these operations will undoubtedly prove helpful.

$$(2,1) + (1,-2) = (3,-1),$$
$$(2,1)\cdot(1,-2) = (2\cdot1 - 1\cdot(-2), 2(-2) + 1\cdot1)$$
$$= (2 + 2, -4 + 1) = (4,-3),$$
$$(1,1) + (-1,-1) = (0,0),$$
$$(1,1)\cdot(1,-1) = (1\cdot1 - 1\cdot(-1), 1(-1) + 1\cdot1)$$
$$= (1 + 1, 0) = (2,0).$$

The complex numbers under the operations of Definition 8.2 satisfy the same properties as the real numbers under the corresponding operations. They are the following, where $(a,b), (c,d)$, and (e,f) are arbitrary complex numbers.

(A1') $(a,b) + (c,d) = (c,d) + (a,b)$ (commutative law).

(A2') $(a,b) + [(c,d) + (e,f)] = [(a,b) + (c,d)] + (e,f)$
(associative law).

(A3') $(a,b) + (0,0) = (a,b)$.

(A4') $(a,b) + (-a,-b) = (0,0)$.

(M1') $(a,b)(c,d) = (c,d)(a,b)$ (commutative law).

(M2') $(a,b)[(c,d)(e,f)] = [(a,b)(c,d)](e,f)$ (associative law).

(M3') $(1,0)(a,b) = (a,b)$.

(M4') $(a,b)\left[\dfrac{a}{a^2 + b^2}, \dfrac{-b}{a^2 + b^2}\right] = (1,0), \qquad a^2 + b^2 \neq 0.$

(D1') $(a,b)[(c,d) + (e,f)] = (a,b)(c,d) + (a,b)(e,f)$ (distributive law).

In addition to these properties, the complex number $(0,1)$ has the property that $(0,1)(0,1) = (0,1)^2 = (-1,0)$. However, $(a,0)(a,0) = (a^2,0)$ for all complex numbers of the form $(a,0)$.

Using the properties of real numbers under addition and multiplication and Definitions 8.1 and 8.2, Properties (A1')–(A4'), (M1')–(M4'), and (D1') can be proved. (A2') and (M1') are proved below.*

Proof of (A2').

$$
\begin{aligned}
(a,b) + [(c,d) + (e,f)] &= (a,b) + (c + e, d + f) \\
&= [a + (c + e), b + (d + f)] \\
&= [(a + c) + e, (b + d) + f] \\
&= (a + c, b + d) + (e,f) \\
&= [(a,b) + (c,d)] + (e,f).
\end{aligned}
$$

Proof of (M1').

$$
\begin{aligned}
(a,b)(c,d) &= (ac - bd, ad + bc) \\
&= (ca - db, cb + da) \\
&= (c,d)(a,b).
\end{aligned}
$$

EXERCISES

1. Perform the indicated operations:
   (a) $(1,2) + (4,1)$
   (b) $(3,-1) + (-2,1)$
   (c) $(\frac{1}{2},1) + (\frac{2}{3},-\frac{3}{4})$
   (d) $(0,1) + (-1,-1)$
   (e) $(3,2)(1,-1)$
   (f) $(0,1)(2,3)$
   (g) $(2,2)(-1,1)$
   (h) $(\frac{1}{2},1)(2,-1)$.
2. Prove (A1'), (M4'), and (D1').

---

*It should be pointed out that, although we essentially took these properties as postulates for the real numbers, their proofs are quite easy for the complex numbers.

2

THE TRIGONOMETRIC FUNCTIONS
AND THEIR PROPERTIES

9. Circular Points

Let

$$C = \{(x,y) : x^2 + y^2 = 1\},$$

where (x,y) is a point in the plane, i.e., C is the circle with center at the origin and having radius one. The arc length of C, or the circumference of the unit circle, is 2π units. The *circular point*, $P(t)$, associated with the real number t is defined as follows.

Definition 9.1. Let t be any real number. Then the circular point $P(t)$ associated with t is the point (x,y) in C which is $|t|$ units from the point $(1,0)$ along C in a counterclockwise direction when $t > 0$ and clockwise when $t < 0$. Hence, $P(t)$ is the end point of an arc of length $|t|$ units with initial point $(1,0)$.

In the terminology of Chapter 1, P is a function whose domain is the set of all real numbers and whose range is the set of points

$$C = \{(x,y) : x^2 + y^2 = 1\}.$$

To note a few values of P, first of all $P(0) = (1,0)$. Since the x and y axes divide C into 4 equal parts, it follows that $P(\pi/2) = (0,1)$, $P(\pi) = (-1,0)$, and $P(3\pi/2) = (0,-1)$.

It is also easy to find the values of P at a few negative numbers as well. For example, by Definition 9.1 it follows that

$$P\left(-\frac{\pi}{2}\right) = (0,-1) = P\left(\frac{3\pi}{2}\right),$$

$$P(-\pi) = (-1,0) = P(\pi),$$

and $$P\left(-\frac{3\pi}{2}\right) = (0,1) = P\left(\frac{\pi}{2}\right).$$

Besides the points indicated above, there are a number of others whose coordinates can be found quite easily, among which are $P(\pi/4)$ and $P(\pi/3)$, which will be obtained below. Table 9.1 in the exercises contains other values of P which should be committed to memory.

Example 9.2. To find the coordinates for $P(\pi/4)$, note that $\pi/4$ is one-half of $\pi/2$, and hence, $P(\pi/4)$ is half-way between $P(0)$ and $P(\pi/2)$. It follows that $P(\pi/4)$ is on the line which bisects the first quadrant. As this line has the equation $y = x$, $P(\pi/4) = (x,x)$. As $P(\pi/4)$ is on C,

$$x^2 + y^2 = 1.$$

But $y = x$ and so $2x^2 = 1$. Therefore, $x^2 = \frac{1}{2}$, and $x = \pm 1/\sqrt{2}$. $P(\pi/4)$ is in the first quadrant where $x > 0$, so $P(\pi/4) = (1/\sqrt{2}, 1/\sqrt{2})$.

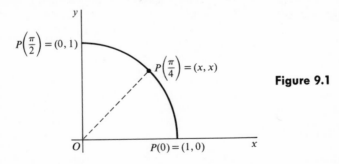

Figure 9.1

Example 9.3.   A slightly more complicated example is the finding of $P(\pi/3)$ using the distance formula.

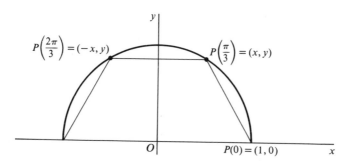

Figure 9.2

Let $P(\pi/3)$ have coordinates (x,y). The points $P(\pi/3)$ and $P(2\pi/3)$ trisect the arc between $P(0)$ and $P(\pi)$, and the point $P(\pi/2)$ bisects it. Thus,

$$\left| P\left(\frac{\pi}{3}\right) P\left(\frac{\pi}{2}\right) \right| = \left| P\left(\frac{\pi}{2}\right) P\left(\frac{2\pi}{3}\right) \right|$$

and it follows that $P(2\pi/3) = (-x,y)$. The distance from $P(0)$ to $P(\pi/3)$ is the same as the distance from $P(\pi/3)$ to $P(2\pi/3)$, i.e.,

$$\left| P(0) P\left(\frac{\pi}{3}\right) \right| = \left| P\left(\frac{\pi}{3}\right) P\left(\frac{2\pi}{3}\right) \right|.$$

In terms of the distance formula,

$$\left| P(0) P\left(\frac{\pi}{3}\right) \right| = \sqrt{(x-1)^2 + y^2}$$

$$\left| P\left(\frac{\pi}{3}\right) P\left(\frac{2\pi}{3}\right) \right| = \sqrt{[x - (-x)]^2 + (y - y)^2} = \sqrt{4x^2} = 2|x|.$$

As these are equal,

$$\sqrt{x^2 - 2x + 1 + y^2} = 2|x| = 2x, \quad \text{for} \quad x > 0.$$

Then

$$x^2 - 2x + 1 + y^2 = 4x^2.$$

But

$$x^2 + y^2 = 1,$$

so

$$4x^2 + 2x - 2 = 0, \quad \text{or} \quad 2x^2 + x - 1 = 0,$$

which factors as

$$(2x - 1)(x + 1) = 0.$$

Hence, the solutions are $x = \frac{1}{2}, -1$. However, $x > 0$, so $x = \frac{1}{2}$ is the solution that is applicable. Then, as

$$y = \sqrt{1 - x^2}, \quad P\left(\frac{\pi}{3}\right) = \left(\frac{1}{2}, \frac{\sqrt{3}}{2}\right).$$

Example 9.4. Using the coordinates of $P(\pi/3)$ it is possible to find the coordinates of $P(\pi/6)$ without using the distance formula, for the length of the arc from $P(0)$ to $P(\pi/6)$ is the same as the length of the arc from $P(\pi/3)$ to $P(\pi/2)$. Hence

$$\left|P(0)P\left(\frac{\pi}{6}\right)\right| = \left|P\left(\frac{\pi}{3}\right)P\left(\frac{\pi}{2}\right)\right|.$$

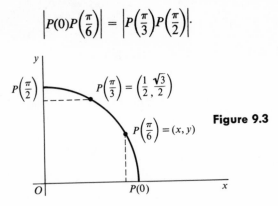

Figure 9.3

It follows that the distance from $P(\pi/6)$ to the x axis, the y coordinate of $P(\pi/6)$, is the same as the distance from $P(\pi/3)$ to the y axis, the x coordinate of $P(\pi/3)$. Then since $x^2 + y^2 = 1$, the x coordinate of $P(\pi/6)$ is the same as the y coordinate of $P(\pi/3)$, and $P(\pi/6) = (\sqrt{3}/2, \frac{1}{2})$.

EXERCISES

Complete the following table:

Table 9.1

t	x coordinates of $P(t)$	y coordinates of $P(t)$
0	1	0
$\dfrac{\pi}{6}$		
$\dfrac{\pi}{4}$	$\dfrac{1}{\sqrt{2}}$	$\dfrac{1}{\sqrt{2}}$

Table 9.1 (*Continued*)

t	x coordinates of $P(t)$	y coordinates of $P(t)$
$\dfrac{\pi}{3}$	$\dfrac{1}{2}$	$\dfrac{\sqrt{3}}{2}$
$\dfrac{\pi}{2}$	0	1
$\dfrac{2\pi}{3}$	$-\dfrac{1}{2}$	$\dfrac{\sqrt{3}}{2}$
$\dfrac{3\pi}{4}$		
$\dfrac{5\pi}{6}$		
π	-1	0
$\dfrac{7\pi}{6}$		
$\dfrac{5\pi}{4}$		
$\dfrac{4\pi}{3}$		
$\dfrac{3\pi}{2}$	0	-1
$\dfrac{5\pi}{3}$		
$\dfrac{7\pi}{4}$		
$\dfrac{11\pi}{6}$		

10. The Trigonometric Functions

The formal definitions of the six trigonometric functions are given below. $P(t) = (x,y)$ is the circular point associated with the real number t.

Definition 10.1. The sine, cosine, tangent, cotangent, secant, and co-secant functions are defined by:

sine $= \{(t,y) : P(t) = (x,y)\}$
cosine $= \{(t,x) : P(t) = (x,y)\}$
tangent $= \{(t,y/x) : P(t) = (x,y),\ x \neq 0\}$
cotangent $= \{(t,x/y) : P(t) = (x,y),\ y \neq 0\}$
secant $= \{(t,1/x) : P(t) = (x,y),\ x \neq 0\}$
cosecant $= \{(t,1/y) : P(t) = (x,y),\ y \neq 0\}$.

Thus if t is any real number and $P(t) = (x,y)$ its circular point, one writes

sine $(t) = y$
cosine $(t) = x$
tangent $(t) = y/x,\ x \neq 0$
cotangent $(t) = x/y,\ y \neq 0$
secant $(t) = 1/x,\ x \neq 0$
cosecant $(t) = 1/y,\ y \neq 0$.

The symbols used to represent the functions are usually abbreviated sin, cos, tan, cot, sec, and csc, respectively. Ordinarily "sin t" is written instead of "sin(t)," also.

From the definition of $P(t)$ and Definition 10.1, it follows that cos 0 = sec 0 = 1, and sin 0 = tan 0 = 0; csc 0 and cot 0 are undefined. Like-wise, as $P(\pi/4) = (1/\sqrt{2}, 1/\sqrt{2})$ it follows that

$$\sin \frac{\pi}{4} = \cos \frac{\pi}{4} = \frac{1}{\sqrt{2}},$$

$$\sec \frac{\pi}{4} = \csc \frac{\pi}{4} = \sqrt{2},$$

and

$$\tan \frac{\pi}{4} = \cot \frac{\pi}{4} = 1.$$

The student should complete the table in the exercises below and learn the entries for these will be used again.

A method is now given which will enable the student to evaluate the functions at any real number t, $0 \leq t \leq 2\pi$, in terms of numbers between 0 and $\pi/2$. In the next section this will be extended to include all real numbers t, not just those between 0 and 2π.

If

$$\frac{\pi}{2} < t < 2\pi$$

and

$$t \neq \pi, t \neq \frac{3\pi}{2},$$

then there are three possible locations for t:

(1) $\quad \frac{\pi}{2} < t < \pi,$

(2) $\quad \pi < t < \frac{3\pi}{2},$

or

(3) $\quad \frac{3\pi}{2} < t < 2\pi.$

These three cases are considered in detail below.

Case 1. $\pi/2 < t < \pi$. Let $t' = \pi - t$. Then $0 < t' < \pi/2$, and if $P(t)$ has coordinates (x,y), it follows that $P(t')$ has coordinates $(-x,y) = (|x|,y)$. For the arc from $P(0)$ to $P(t')$ is the same length as the arc from $P(\pi)$ to $P(t)$.

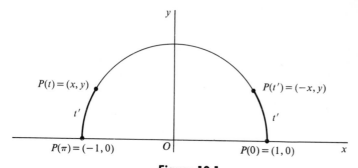

Figure 10.1

Now if T is any of the trigonometric functions, then $|T(t)| = T(t')$. Hence, $T(t)$ and $T(t')$ differ at most in sign, and the appropriate sign is determined from Problem 2 below.

Case 2. $\pi < t < 3\pi/2$. If $t' = t - \pi$, then $0 < t' < \pi/2$. If $P(t)$ has coordinates (x,y), then $P(t')$ has coordinates $(-x,-y) = (|x|,|y|)$, for the arc from $P(0)$ to $P(t')$ is the same length as the arc from $P(\pi)$ to $P(t)$.

If T is any trigonometric function, $|T(t)| = T(t')$. Hence, $T(t)$ and $T(t')$ differ at most in sign, and this is determined from Problem 2 below.

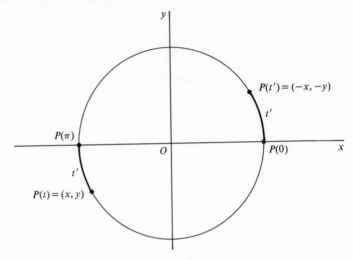

Figure 10.2

Case 3. $3\pi/2 < t < 2\pi$. If $t' = 2\pi - t$, then $0 < t' < \pi/2$. If $P(t)$ has coordinates (x,y) then $P(t')$ has coordinates $(x,-y) = (x,|y|)$ for reasons similar to those stated previously in Cases 1 and 2.

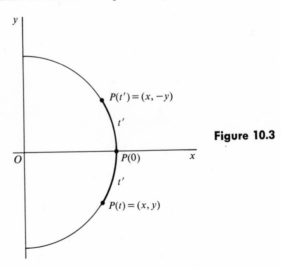

Figure 10.3

If T is any trigonometric function, then $|T(t)| = T(t')$ as before, and the appropriate sign difference between $T(t)$ and $T(t')$ may again be determined from Problem 2 below.

In each of these three cases, given t and its circular point $P(t)$ there arises a number t', $0 < t' < \pi/2$, such that $|T(t)| = T(t')$. This gives rise to the following definition.

Definition 10.2. If $\pi/2 < t < 2\pi$ and $t \neq \pi$, $3\pi/2$, then one and only one of the three numbers $\pi - t$, $t - \pi$, and $2\pi - t$ is between 0 and $\pi/2$. This number is called the *reference number* associated with t. If t' is the reference number of t and T any trigonometric function, then $T(t') = |T(t)|$.

Some examples using this definition will perhaps be helpful.

Example 10.3. The reference number associated with $7\pi/12$ is $\pi - 7\pi/12 = 5\pi/12$. As $P(7\pi/12)$ is in the second quadrant,

$$\sin \frac{7\pi}{12} = \sin \frac{5\pi}{12}, \qquad \cos \frac{7\pi}{12} = -\cos \frac{5\pi}{12}, \qquad \tan \frac{7\pi}{12} = -\tan \frac{5\pi}{12},$$

$$\cot \frac{7\pi}{12} = -\cot \frac{5\pi}{12}, \qquad \sec \frac{7\pi}{12} = -\sec \frac{5\pi}{12} \quad \text{and} \quad \csc \frac{7\pi}{12} = \csc \frac{5\pi}{12}.$$

Example 10.4. Using 3.14 as an approximation for π, the reference number of 3.5 is approximately $0.36 = 3.5 - 3.14$. Using "$\doteq$" to represent approximately equal, $|T(3.5)| \doteq T(0.36)$ for any trigonometric function T. (The tangent and cotangent are positive and the others are negative.) For example, from Table II (page 160), $\sin 0.36 \doteq 0.3523$, so $\sin 3.5 \doteq -0.3523$.

EXERCISES

1. Complete and learn Table 10.1. Use Table 9.1 to find the values between 0 and $\pi/2$ and use Definition 10.2 to fill in the remainder.

2. Determine the signs of the trigonometric functions for $0 < t < \pi/2$, $\pi/2 < t < \pi$, $\pi < t < 3\pi/2$, and $3\pi/2 < t < 2\pi$.

3. Write each of the following in terms of the same function evaluated between 0 and $\pi/2$.

   (a) $\sin 9\pi/5$ (c) $\tan 13\pi/7$ (e) $\sec 14\pi/15$

   (b) $\cos 8\pi/11$ (d) $\cot 17\pi/12$ (f) $\csc 14\pi/13$.

Table 10.1

t	sin t	cos t	tan t	cot t	sec t	csc t
0	0	1	0	undefined	1	undefined
$\dfrac{\pi}{6}$						
$\dfrac{\pi}{4}$						
$\dfrac{\pi}{3}$						
$\dfrac{\pi}{2}$	1	0	undefined	0	undefined	1
$\dfrac{2\pi}{3}$						
$\dfrac{3\pi}{4}$						
$\dfrac{5\pi}{6}$						
π	0	-1	0	undefined	-1	undefined
$\dfrac{7\pi}{6}$						
$\dfrac{5\pi}{4}$						
$\dfrac{4\pi}{3}$						
$\dfrac{3\pi}{2}$	-1	0	undefined	0	undefined	-1
$\dfrac{5\pi}{3}$						
$\dfrac{7\pi}{4}$						
$\dfrac{11\pi}{6}$						
2π						

4. Use 3.14 as an approximation for π, Definition 10.2, and Table II to find the following:

(a) sin 2   (d) sec 6   (g) sin 4.25
(b) cos 3   (e) cot 5   (h) cos 5.65
(c) tan 4.5   (f) csc 3.2   (j) tan 5.75.

5. A function f is said to be *increasing* in an interval, if and only if $f(x_1) < f(x_2)$, when $x_1 < x_2$ in the interval under consideration. f is *decreasing* in an interval, if and only if $f(x_2) < f(x_1)$, when $x_1 < x_2$ in the interval under consideration. Use the unit circle to determine where the sine and cosine functions are increasing and where they are decreasing.

11. Periodic Functions

The definition of the circular point gives rise to a phenomenon that occurs frequently in mathematics. For any real number t, the numbers t and $t + 2\pi$ are 2π units apart. As the circumference of C is 2π, it follows that $P(t) = P(t + 2\pi)$. From this observation and Definition 10.1, it follows that for any trigonometric function T, $T(t + 2\pi) = T(t)$ for every real number t. Intuitively the values of the functions are "repeated at regular intervals."

Now consider another example of a function which has this property of repeating its values. Let f be defined by $f(x) = x - [x]$, where $[x]$ is the greatest integer $\leq x$. (See Example 7.4.) For $0 \leq x < 1$, $f(x) = x$; for $1 \leq x < 2$, $f(x) = x - 1$; for $2 \leq x < 3$, $f(x) = x - 2$. In general, for $n \leq x < n + 1$, $f(x) = x - n$, n any integer. Then $f(0) = f(1) = f(2) = \cdots = f(n) = 0$ for any integer n. In general, if x is any real number, then there is an integer n such that $n \leq x < n + 1$ and $f(x) = x - n$. But then $n + 1 \leq x + 1 < n + 2$ and $f(x + 1) = (x + 1) - (n + 1) = x - n$. Thus, $f(x + 1) = f(x)$ for all real numbers x. The graph of f is given below.

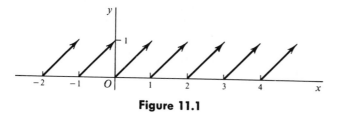

Figure 11.1

An arrow has been used here to indicate that $f(x) \neq 1$ for all x. In each of the examples considered above, there is a positive number

p such that $f(x + p) = f(x)$ for all x in the domain of f. In each instance there is a smallest such positive number p. Thus the following definition is obtained.

Definition 11.1. Let f be a function whose domain is a set of real numbers. f is said to be *periodic*, if and only if there is a $p > 0$ such that $f(x + p) = f(x)$ for all x in the domain of f. If f is periodic and there is a smallest such number p, it is called the *primitive period* or, when there is no cause for confusion, simply the period of f.

Note. There are functions which are periodic but have no primitive period. For example, if f is a constant function defined on the set of real numbers, then $f(x + p) = f(x)$ for all $p > 0$ and for all x.

As was observed above, $P(t + 2\pi) = P(t)$ for all t, where $P(t)$ is the circular point. However, if $0 < p < 2\pi$, then $P(0) \neq P(p)$, so P is periodic with primitive period 2π. As $P(t + 2\pi) = P(t)$ for all t, it follows that $T(t + 2\pi) = T(t)$ for any trigonometric function T and for all t for which T is defined. As $\cos 0 = \sec 0 = 1$, whereas $\cos t \neq 1$ and $\sec t \neq 1$ for $0 < t < 2\pi$, and $\sin \pi/2 = \csc \pi/2 = 1$ whereas $\sin t \neq 1$ and $\csc t \neq 1$ for $0 < t < 2\pi$, $t \neq \pi/2$, it follows that these four trigonometric functions have primitive period 2π. The tangent and cotangent are also periodic, but, as $P(t + \pi) = (-x, -y)$ when $P(t) = (x, y)$, it follows that $\tan (t + \pi) = \tan t$ and $\cot (t + \pi) = \cot t$ for all real numbers t in the respective domains. Hence, they have primitive period π.

The other function given above, defined by $f(x) = x - [x]$, is periodic with primitive period 1. Another way of obtaining this function is to define a function F by $F(x) = x$, $0 \leq x < 1$. Then "extend" its domain to include all real numbers by writing

$$f(x) = F(x), \quad 0 \leq x < 1, \quad \text{and}$$

$$f(x + 1) = f(x) \quad \text{for all real } x.$$

Another simple example of a periodic function comes from defining a function g by letting

$$g(x) = \begin{cases} 1 & \text{if } 0 \leq x < 1, \\ -1 & \text{if } 1 \leq x < 2, \end{cases}$$

and
$$g(x + 2) = g(x) \quad \text{for all real } x.$$

The graph of g for $-2 \leq x \leq 3$ is given below.

Figure 11.2

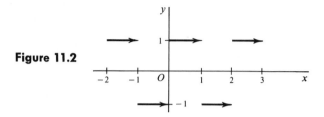

As these examples illustrate, the trigonometric functions are not the only functions which are periodic. Theorem 11.2 below gives a method for constructing a large class of periodic functions, starting with only a few.

Theorem 11.2.   Let g be a nonconstant function and f a periodic function with primitive period p. Then the function h defined by $h(x) = g[f(x)]$, i.e., $h = g \circ f$, is periodic with primitive period $\leq p$.

Proof.   Let x be any number in the domain of h. Then $h(x + p) = g[f(x + p)] = g[f(x)] = h(x)$ and h is periodic. The period of h will in general depend on g as well as p.

If f is periodic and g is nonconstant, then the function defined by $f[g(x)]$ is not periodic in general. However, we are able to prove the following theorem.

Theorem 11.3.   If f is periodic with primitive period p and h is defined by $h(x) = f(ax)$ for $a > 0$, then h is periodic with primitive period p/a.

Note.   This can be thought of as $h(x) = f[g(x)]$, where f has primitive period p and $g(x) = ax$.

Proof.   As f is periodic with primitive period p, we have $f(ax + p) = f(ax)$ for all (ax) in the domain of f. Hence, h is periodic. Now we observe that

$$f(ax + p) = f\left[a\left(x + \frac{p}{a}\right)\right] = h\left(x + \frac{p}{a}\right) = h(x).$$

As p is the smallest positive number such that $f(x + p) = f(x)$ for all x in domain f, it follows that p/a is the smallest number q such that $h(x + q) = h(x)$ for all x in domain h. Thus the primitive period of h is p/a.

This theorem has many applications. For example, if f is the sine function and $h(x) = f(2x) = \sin 2x$, then h has primitive period $p = 2\pi/2$ as the primitive period of f is 2π. If f is the sine function and $h(x) = f(\frac{3}{4}x) = \sin \frac{3}{4}x$, then h has primitive period

$$p = \frac{2\pi}{\frac{3}{4}} = \frac{8\pi}{3}.$$

Another important theorem involving periodic functions arises by observing that, for the circular point function P, we have $P(t) = P[t + n(2\pi)]$ for all t and all integers n—positive, negative, or zero. This is characteristic of all periodic functions and is the content of the following theorem.

Theorem 11.4. Let f be periodic with primitive period p. Then

$$f(t) = f(t \pm np)$$

for every positive integer n and for all t in the domain of f.

Proof. The proof is done by mathematical induction. By definition, $f(\tau + p) = f(\tau)$ for every τ in the domain of f. Hence, let t be an arbitrary number in the domain of f, and let $\tau = t - p$. Then $f(\tau) = f(t - p) = f(\tau + p) = f(t)$. Hence, $f(t \pm p) = f(t)$.

Assume now that the theorem is true for the integer n. Then $f(t \pm np) = f(t)$. $f[t + (n + 1)p] = f[(t + np) + p] = f(t + np)$, and $f[t - (n + 1)p] = f[(t - np) - p] = f(t - np)$. As $f(t \pm np) = f(t)$, it follows that $f[t \pm (n + 1)p] = f(t)$. Thus the theorem is true for $n = 1$, and we have shown that, if it is true for n, it is true for $n + 1$. Therefore, by mathematical induction, it is true for all positive integers n.

As any real number t can be written in the form $t = k \cdot 2\pi + r$, where k is some integer and $0 \le r < 2\pi$, we are able to evaluate the trigonometric functions at any real number using only values between 0 and 2π and Theorem 11.4.
Thus

$$\sin \frac{43\pi}{4} = \sin \frac{3\pi}{4},$$

as

$$\frac{43\pi}{4} = 10\pi + \frac{3\pi}{4} = 5 \cdot 2\pi + \frac{3\pi}{4}.$$

Also,

$$\sin\left(\frac{-9\pi}{4}\right) = \sin\left(\frac{-16\pi}{4} + \frac{7\pi}{4}\right) = \sin\frac{7\pi}{4}.$$

We can, of course, carry the work even further using reference numbers. Thus

$$\sin\frac{43\pi}{4} = \sin\frac{3\pi}{4} = \sin\frac{\pi}{4},$$

and

$$\sin\left(\frac{-9\pi}{4}\right) = \sin\frac{7\pi}{4} = -\sin\frac{\pi}{4}.$$

EXERCISES

1. Evaluate each of the following:

(a) $\sin\frac{27\pi}{4}$

(b) $\cos\frac{229\pi}{3}$

(c) $\tan\frac{23\pi}{6}$

(d) $\sec\left(\frac{-27\pi}{4}\right)$

(e) $\csc\frac{77\pi}{6}$

(f) $\cot\frac{47\pi}{3}$

(g) $\sin\left(\frac{-19\pi}{3}\right)$

(h) $\cos\left(\frac{-47\pi}{4}\right)$

(i) $\tan\left(\frac{-35\pi}{4}\right)$.

2. Using 3.14 as an approximation for π, Theorem 11.4, and Table II (page 160), evaluate each of the following.

(a) $\sin 21$

(b) $\cos(-12)$

(c) $\sin(-30)$

(d) $\tan 20$

(e) $\cot(-10.5)$

(f) $\cos 100$

(g) $\sin 70$

(h) $\cos 29$

(i) $\tan(-25)$.

3. Use Theorem 11.3 to find the period of each of the following functions f, defined by:

(a) $f(x) = \sin 2x$

(b) $f(x) = \cos\frac{2x}{3}$

(c) $f(x) = \tan 2x$

(d) $f(x) = \cot\frac{x}{4}$

(e) $f(x) = \frac{x}{2} - \left[\frac{x}{2}\right]$

(f) $f(x) = 3x - [3x]$

(g) $f(x) = \sin 2\pi x$

(h) $f(x) = \cos\frac{2x}{\pi}$

(i) $f(x) = \tan\frac{5\pi x}{4}$.

12. Fundamental Identities

A number of equations relating the trigonometric functions arise from the definitions of the trigonometric functions and the equation $x^2 + y^2 = 1$. We list some examples on the following page.

Example 12.1. If t is any real number, then $P(t) = (x,y) = (\cos t, \sin t)$. As $x^2 + y^2 = 1$, it follows that

$$\cos^2 t + \sin^2 t = 1*$$

for every real number t.

Example 12.2. In the definitions of $\sin t$ and $\csc t$, given $P(t) = (x,y)$, $y = \sin t$ and $\csc t = 1/y$, $y \neq 0$. Thus,

$$\csc t = \frac{1}{\sin t}$$

for all t for which $\sin t \neq 0$.

Example 12.3. The equation

$$1 + \tan^2 t = \sec^2 t$$

is true for all t for which $\tan t$ and $\sec t$ are defined. When $x \neq 0$, we can write the sequence:

$$x^2 + y^2 = 1$$

$$1 + \frac{y^2}{x^2} = \frac{1}{x^2}$$

$$1 + \left(\frac{y}{x}\right)^2 = \left(\frac{1}{x}\right)^2$$

$$1 + (\tan t)^2 = (\sec t)^2$$

or $$1 + \tan^2 t = \sec^2 t.$$

Here, as before, $$P(t) = (x,y).$$

In all three of these examples the quantity on the left-hand side of the equation is equal to the quantity on the right-hand side for all t, for which both are defined. In this respect, these equations are undoubtedly different from most of the equations the student has encountered. For example, the quantity $x + 1$ equals the quantity $2x - 1$, $x + 1 =$

---

*The exponent appears in this position except for -1. Thus we write $1/\tan t$ = $(\tan t)^{-1}$, because $\tan^{-1} t$ is given a very different meaning.

$2x - 1$, only if $x = 2$. We are led to distinguish between these two types of equations in the definition below.

Definition 12.4.   An equation which is valid for all numbers for which the two sides are defined is called an *identical equation* or an *identity*. An equation which is not valid for at least one number at which both sides are defined is called a *conditional equation*, or just an *equation*.

From the definition it follows that the equations of Examples 12.1–12.3 are identities while the equation $x + 1 = 2x - 1$ is a conditional equation.

We now list some fundamental identities whose derivations follow from the definitions of the trigonometric functions and the equation $x^2 + y^2 = 1$, where $(x,y) = P(t)$. Those which have not already been derived are left as exercises. All should be committed to memory.

(1) $\sin^2 t + \cos^2 t = 1.$ (5) $\sec t = \dfrac{1}{\cos t}.$

(2) $1 + \tan^2 t = \sec^2 t.$ (6) $\cot t = \dfrac{1}{\tan t}.$

(3) $1 + \cot^2 t = \csc^2 t.$ (7) $\tan t = \dfrac{\sin t}{\cos t}.$

(4) $\csc t = \dfrac{1}{\sin t}.$ (8) $\cot t = \dfrac{\cos t}{\sin t}.$

The difference between identities and conditional equations should suggest that they are handled in an entirely different manner. Thus, we "solve" a conditional equation and "prove" or "verify" an identity. To solve an equation is to find all the numbers which satisfy it. To verify an identity means to show that a given equation is an identity. In either case it is essential that the operations involved are valid and reversible. For example, multiplication or division by zero is excluded.

We will illustrate below two ways of attacking the same identity.

Example 12.5.   Verify that

$$\frac{1}{\cos t} - \frac{\sin^2 t}{\cos t} = \cos t$$

is an identity.

Solution 1.   As

$$\frac{1}{\cos t} - \frac{\sin^2 t}{\cos t}$$

is not defined for cos $t = 0$, those values of t are excluded. Multiply both sides of the equation by cos t. Then we have

(12.1) $$1 - \sin^2 t = \cos^2 t.$$

By Identity (1), above, we see that $1 - \sin^2 t = \cos^2 t$ is an identity. However, this is not the identity we wished to verify, and the verification is not complete unless both sides of the equation are now divided by cos t.

Solution 1 is technically correct but not very desirable, as most students would stop after obtaining the equation (12.1). A better approach appears to be a "reduction" of the left-hand side to the right-hand side as follows.

Solution 2.

(a) $$\frac{1}{\cos t} - \frac{\sin^2 t}{\cos t} = \frac{1 - \sin^2 t}{\cos t}$$

(b) $$= \frac{\cos^2 t}{\cos t}$$

(c) $$= \cos t \ (\cos t \neq 0).$$

Step (a) is obtained by performing the operation of subtracting the second fraction from the first; step (b) is the replacement of one expression by an equivalent one using an identity which has already been obtained; and step (c) is the cancellation of a nonzero term from the numerator and denominator of a fraction. All of the operations used in this solution are clearly reversible.

The author prefers the method of reduction as illustrated in Solution 2. We now list four steps to be followed in performing a reduction to prove an equation is an identity.

(1) Choose the side of the equation on which the operations are to be performed, usually the more complicated expression.
(2) Perform any indicated operations, such as addition, subtraction, multiplication, or division.
(3) Simplify the expression by making use of any previously obtained identities.
(4) If other attempts have failed to yield the result, write the expression entirely in terms of sines and cosines and then proceed.

We hasten to point out that steps 2 and 3 may easily be interchanged. The most important thing is to begin with step 1.

We now give more examples to illustrate the above statements.

Example 12.6.   Show that the equation

$$\frac{\sin x}{\cos x} + \frac{\cos x}{\sin x} = \sec x \csc x$$

is an identity.

Solution.   The left side of the equation is clearly the one to be reduced. The series of steps used in reducing the left side to the right side is listed below, together with a brief explanation of each.

$$\frac{\sin x}{\cos x} + \frac{\cos x}{\sin x} = \frac{\sin^2 x + \cos^2 x}{\cos x \sin x}$$ Addition of fractions.

$$= \frac{1}{\cos x \sin x}$$ Replacement of equivalent quantity using identity.

$$= \frac{1}{\cos x} \frac{1}{\sin x}$$ Multiplication of fractions.

$$= \sec x \csc x$$ Replacement by equivalent quantity using identity.

Example 12.7.   Show that

$$\frac{\sin x}{1 + \cos x} = \frac{1 - \cos x}{\sin x}$$

is an identity.

Solution.   Here, it makes no difference which side we choose to reduce, so we arbitrarily choose the left side, and derive the right side from it. We may write:

$$\frac{\sin x}{1 + \cos x} = \frac{\sin x \,(1 - \cos x)}{(1 + \cos x)\,(1 - \cos x)}, \qquad 1 - \cos x \neq 0$$

$$= \frac{\sin x \,(1 - \cos x)}{1 - \cos^2 x}$$

$$= \frac{\sin x \,(1 - \cos x)}{\sin^2 x}$$

$$= \frac{1 - \cos x}{\sin x}.$$

The reasons for, and the validity of the separate statements should be clear except, perhaps, for the first one. Why multiply and divide by $1 - \cos x$? This is a perfectly valid and reversible operation so long as $1 - \cos x \neq 0$, and the reason for making it is seen if we look carefully at what we are trying to derive:

$$\frac{1 - \cos x}{\sin x}.$$

The numerator of this fraction is $1 - \cos x$, and we must obtain this from $\sin x$, the numerator of the original fraction. One way to do this is to proceed in precisely the manner we have indicated.

Trigonometric identities have many uses, one of which we will now illustrate.

Example 12.8. Given $\sin t_0 = \frac{3}{5}$ and $\cos t_0 < 0$, find the values of the other five trigonometric functions at t_0.

Solution. As $\sin t_0 = \frac{3}{5}$ and $\csc t_0 = 1/\sin t_0$, it follows that $\csc t_0 = \frac{5}{3}$. To find the other values we need only find $\cos t_0$. As $\sin^2 t + \cos^2 t = 1$ for all real numbers t, then in particular $\cos^2 t_0 + \sin^2 t_0 = 1$ or $\cos^2 t_0 = 1 - \sin^2 t_0 = 1 - 9/25 = 16/25$. Thus $|\cos t_0| = \frac{4}{5}$ and as $\cos t_0 < 0$, it follows that $\cos t_0 = -\frac{4}{5}$. Then

$$\sec t_0 = \frac{1}{\cos t_0} = -\tfrac{5}{4},$$

$$\tan t_0 = \frac{\sin t_0}{\cos t_0} = \frac{\frac{3}{5}}{-\frac{4}{5}} = -\tfrac{3}{4},$$

and
$$\cot t_0 = \frac{1}{\tan t_0} = -\tfrac{4}{3}.$$

EXERCISES

1. Derive the identities (3), (5), (6), (7), and (8) of this section.
2. Given $\sin t = \frac{2}{3}$, find possible values of the other trigonometric functions at t.
3. Given $\cos t = -7/25$ and $\sin t > 0$, find the values of the other trigonometric functions at t.
4. Given $\tan t = \frac{3}{4}$, find $\sin t$ and $\cos t$ if $\sin t < 0$.
5. Given $\tan t = \frac{5}{7}$, find $\sin t$ and $\cos t$ if $\sin t > 0$.

6. Given $\cot t = 5$, find $\sin t$ and $\cos t$ if $\sin t > 0$.

Use reduction on one side of each equation to prove that it is an identity.

7. $\tan t + \cot t = \sec t \csc t$.

10. $\dfrac{1 - \cos x}{1 + \cos x} = (\csc x - \cot x)^2$.

8. $\sec^2 t + \csc^2 t = \sec^2 t \csc^2 t$.

11. $\dfrac{\csc^2 x - 1}{\sec^2 x - 1} = \cot^4 x$.

9. $\dfrac{1 - \sin^2 x}{\sin^2 x} = \cot^2 x$.

12. $\dfrac{\sin t}{\sec t} = \dfrac{1}{\tan t + \cot t}$.

13. $\csc^4 x - \cot^4 x = \csc^2 x + \cot^2 x$.

14. $(\tan t + \cot t)^2 = \sec^2 t + \csc^2 t$.

15. $(1 + \sec t)(1 - \cos t) = \dfrac{\sec t}{\csc^2 t}$.

16. $\cos^4 t - \sin^4 t = 1 - 2 \sin^2 t$.

17. $\dfrac{1}{1 - \cos x} + \dfrac{1}{1 + \cos x} = 2(1 + \cot^2 x)$.

18. $\dfrac{1 + \sin^2 t \sec^2 t}{1 + \cos^2 t \csc^2 t} = \tan^2 t$.

19. $(1 - \cos x)(1 + \sec x)(1 - \sin x)(1 + \csc x) = \sin x \cos x$.

20. $\dfrac{\tan t + \sin t}{\csc t + \cot t} = \sin^2 t \sec t$.

21. $\cot t \csc t = \dfrac{1}{\sec t - \cos t}$.

22. $\tan^2 t \sin^2 t = \tan^2 t - \sin^2 t$.

23. $\dfrac{\sin t}{1 + \cos t} + \dfrac{\cos t}{1 - \cos t} = 2 \csc t$.

24. $\dfrac{1}{1 - \sin t} + \dfrac{1}{1 + \sin t} = 2(1 + \tan^2 t)$.

25. $\dfrac{\cot t - \tan t}{\tan t + \cot t} = 1 - 2 \sin^2 t$.

26. $(\tan t + \cot t)^2 \sin^2 t - \tan^2 t = 1$.

27. $\dfrac{\sec x - \csc x}{\sec x + \csc x} = \dfrac{\tan x - 1}{\tan x + 1}$.

28. $\dfrac{\tan x - \sin x}{\sin^3 x} = \dfrac{\sec x}{1 + \cos x}$.

29. $\dfrac{1 - 2 \cos^2 x}{\sin x \cos x} = \tan x - \cot x$.

30. $\dfrac{1}{1 + \sin^2 t} + \dfrac{1}{1 + \cos^2 t} + \dfrac{1}{1 + \sec^2 t} + \dfrac{1}{1 + \csc^2 t} = 2$.

13. The Derivation of a Formula for cos (*t* − *s*) for Arbitrary Real Numbers *t* and s

Let A and B be two real numbers with $P(A)$, $P(B)$, and $P(A + B)$ the circular points of A, B and $A + B$. From the definitions of $P(t)$ and the trigonometric functions, it follows that we may write $P(t) = (x,y) = (\cos t, \sin t)$ for each real number t. Thus $P(A) = (\cos A, \sin A)$, $P(B) =$

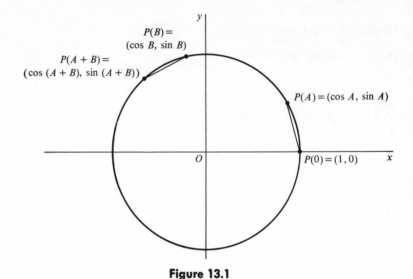

Figure 13.1

$(\cos B, \sin B)$, and $P(A + B) = (\cos (A + B), \sin (A + B))$. (See Figure 13.1.) Furthermore, the arc from $P(B)$ to $P(A + B)$ is equal to the arc from $P(0)$ to $P(A)$. Thus,

$$|P(0)P(A)| = |P(B)P(A + B)|,$$

or

$$\sqrt{(1 - \cos A)^2 + \sin^2 A} = \sqrt{[\cos B - \cos (A + B)]^2 + [\sin B - \sin (A + B)]^2}.$$

Then squaring both sides gives

$$1 - 2 \cos A + \cos^2 A + \sin^2 A = \cos^2 B - 2 \cos B \cos (A + B) + \cos^2 (A + B) + \sin^2 B - 2 \sin B \sin (A + B) + \sin^2 (A + B).$$

When we collect terms and simplify, we have

(13.1) $\cos A = \cos (A + B) \cos B + \sin (A + B) \sin B.$

This equation is valid for all real numbers A and B. Thus, if s and t are any real numbers, let $A = t - s$ and $B = s$ in Equation (13.1) to obtain

(13.2) $\cos (t - s) = \cos t \cos s + \sin t \sin s.$

This is the basic formula that will be used in deriving most of the formulas involving functions evaluated at sums and differences.

14. Formulas for cos $(t + s)$, sin $(t \pm s)$, and tan $(t \pm s)$

In formula (13.2) above, let $t = \pi/2$ and s be an arbitrary real number. Then

$$\cos \left(\frac{\pi}{2} - s\right) = \cos \frac{\pi}{2} \cos s + \sin \frac{\pi}{2} \sin s.$$

But

$$\cos \frac{\pi}{2} = 0 \quad \text{and} \quad \sin \frac{\pi}{2} = 1.$$

Hence,

(14.1) $$\cos \left(\frac{\pi}{2} - s\right) = \sin s$$

for every real number s. Now if $s = \pi/2 - t$ in Equation (14.1), we have

$$\sin \left(\frac{\pi}{2} - t\right) = \cos \left[\frac{\pi}{2} - \left(\frac{\pi}{2} - t\right)\right]$$

$$= \cos \left(\frac{\pi}{2} - \frac{\pi}{2} + t\right) = \cos t.$$

Hence, for every real number t,

(14.2) $$\sin \left(\frac{\pi}{2} - t\right) = \cos t.$$

Equations (14.1) and (14.2) are reasons for saying that the sine and cosine functions are "co-functions." The reader will show similar relations between the other two pairs of functions in the exercises. The chief use here for these two equations is in developing other formulas.

Now in Equation (13.2), let $t = 0$ and s be arbitrary. Then

$$\cos(-s) = \cos(0 - s) = \cos 0 \cos s + \sin 0 \sin s.$$

As $\cos 0 = 1$ and $\sin 0 = 0$, it follows that

(14.3) $$\cos(-s) = \cos s$$

for every real number s. A formula for $\sin(-t)$ is obtained by observing that in Equation (14.1),

$$\sin(-t) = \cos\left(\frac{\pi}{2} + t\right).$$

By Equation (14.3),

$$\cos\left(\frac{\pi}{2} + t\right) = \cos\left(-\frac{\pi}{2} - t\right),$$

and by Equation (13.2)

$$\cos\left(-\frac{\pi}{2} - t\right) = \cos\left(-\frac{\pi}{2}\right)\cos t + \sin\left(-\frac{\pi}{2}\right)\sin t.$$

But $\cos\left(-\frac{\pi}{2}\right) = 0$ and $\sin\left(-\frac{\pi}{2}\right) = -1$. Thus, as $\sin(-t) = \cos\left(-\frac{\pi}{2} - t\right)$, it follows that

(14.4) $$\sin(-t) = -\sin t$$

for every real number t.

We are now prepared to give easy derivations of $\cos(t + s)$ and $\sin(t \pm s)$. As

$$\cos(t + s) = \cos(t - (-s))$$
$$= \cos t \cos(-s) + \sin t \sin(-s),$$

it follows by Equations (14.3) and (14.4) that

(14.5) $$\cos(t + s) = \cos t \cos s - \sin t \sin s.$$

To develop a formula for $\sin (t + s)$, write

$$\sin (t + s) = \cos \left[\frac{\pi}{2} - (t + s) \right]$$

$$= \cos \left[\left(\frac{\pi}{2} - t \right) - s \right]$$

$$= \cos \left(\frac{\pi}{2} - t \right) \cos s + \sin \left(\frac{\pi}{2} - t \right) \sin s.$$

By Equations (14.1) and (14.2), we have

(14.6) $\sin (t + s) = \sin t \cos s + \cos t \sin s.$

As $\sin (t - s) = \sin [t + (-s)]$, it follows immediately that

(14.7) $\sin (t - s) = \sin t \cos s - \cos t \sin s.$

The formulas for $\tan (t \pm s)$ can be obtained simultaneously as follows:

$$\tan (t \pm s) = \frac{\sin (t \pm s)}{\cos (t \pm s)}$$

$$= \frac{\sin t \cos s \pm \cos t \sin s}{\cos t \cos s \mp \sin t \sin s}$$

$$= \frac{\dfrac{\sin t \cos s \pm \cos t \sin s}{\cos t \cos s}}{\dfrac{\cos t \cos s \mp \sin t \sin s}{\cos t \cos s}}$$

$$= \frac{\dfrac{\sin t}{\cos t} \pm \dfrac{\sin s}{\cos s}}{1 \mp \dfrac{\sin t \sin s}{\cos t \cos s}}.$$

Hence, we have

(14.8) $$\tan (t \pm s) = \frac{\tan t \pm \tan s}{1 \mp \tan t \tan s},$$

provided $\tan t$, $\tan s$ and $\tan (t \pm s)$ are defined.

For convenience we list below the formulas derived in sections **13** and **14.**

$$\cos (t + s) = \cos t \cos s - \sin t \sin s.$$

$$\cos (t - s) = \cos t \cos s + \sin t \sin s.$$

$$\sin (t + s) = \sin t \cos s + \cos t \sin s.$$

$$\sin (t - s) = \sin t \cos s - \cos t \sin s.$$

$$\tan (t + s) = \frac{\tan t + \tan s}{1 - \tan t \tan s}.$$

$$\tan (t - s) = \frac{\tan t - \tan s}{1 + \tan t \tan s}.$$

$$\sin \left(\frac{\pi}{2} - t\right) = \cos t.$$

$$\cos \left(\frac{\pi}{2} - t\right) = \sin t.$$

$$\sin (- t) = -\sin t$$

$$\cos (- t) = \cos t.$$

The formulas listed above are identities. Together with entries of Table 10.1 they can be used to find such values as $\cos \pi/12$, $\tan 5\pi/12$, and so on. As an example, since $\pi/12 = \pi/3 - \pi/4$, it follows that

$$\cos \frac{\pi}{12} = \cos \left(\frac{\pi}{3} - \frac{\pi}{4}\right) = \cos \frac{\pi}{3} \cos \frac{\pi}{4} + \sin \frac{\pi}{3} \sin \frac{\pi}{4}$$

$$= \frac{1}{2} \frac{\sqrt{2}}{2} + \frac{\sqrt{3}}{2} \frac{\sqrt{2}}{2} = \frac{\sqrt{2}}{4} (1 + \sqrt{3}).$$

EXERCISES

1. Derive formulas for $\cot (t \pm s)$ in terms of $\cot t$ and $\cot s$.
2. If $\sin t = \frac{3}{4}$, $\cos s = \frac{2}{3}$, $\tan t < 0$, and $\tan s > 0$, find $\sin (s + t)$ and $\cos (s - t)$.
3. If $\tan t = 5/12$, $\cot s = \frac{3}{4}$, $\sin t > 0$ and $\cos s < 0$, find $\sin (t + s)$, $\cos (t - s)$, and $\tan (t + s)$.
4. Use identities from above and Table 10.1 to evaluate each of the following:

(a) $\tan \dfrac{5\pi}{12}$ (c) $\sin \dfrac{7\pi}{12}$ (e) $\tan \dfrac{\pi}{12}$

(b) $\cos \dfrac{7\pi}{12}$ (d) $\cos \dfrac{13\pi}{12}$ (f) $\sin \dfrac{5\pi}{12}.$

5. Show that for any real number t,

   (a) $\cos(\pi + t) = -\cos t$ (d) $\sin(\pi - t) = \sin t$

   (b) $\cos(\pi - t) = -\cos t$ (e) $\sin\left(\dfrac{3\pi}{2} + t\right) = -\cos t$

   (c) $\sin(\pi + t) = -\sin t$ (f) $\cos\left(\dfrac{3\pi}{2} + t\right) = \sin t.$

Prove the following identities using methods similar to those of section **12**:

6. $\tan\left(\dfrac{\pi}{4} + t\right) = \dfrac{1 + \tan t}{1 - \tan t}.$

7. $\sin(t + s) + \sin(t - s) = 2\sin t \cos s.$

8. $\sin(t + s)\sin(t - s) = \sin^2 t - \sin^2 s.$

9. $\cos(t + s) + \cos(t - s) = 2\cos t \cos s.$

10. $2\sin\left(t + \dfrac{\pi}{4}\right)\sin\left(t - \dfrac{\pi}{4}\right) = \sin^2 t - \cos^2 t.$

11. $\dfrac{\sin(t + s)}{\sin(t - s)} = \dfrac{\tan t + \tan s}{\tan t - \tan s}.$

12. $\cot\left(\dfrac{\pi}{4} - t\right) = \dfrac{\cot t + 1}{\cot t - 1}.$

13. $\sin\left(\dfrac{\pi}{4} + t\right) = \cos\left(\dfrac{\pi}{4} - t\right).$

14. $\sin(t + s)\sin s + \cos(t + s)\cos s = \cos t.$

15. $\cos(t + s)\cos(t - s) = \cos^2 t - \sin^2 s.$

16. $\dfrac{\sin 2x}{\sin x} - \dfrac{\cos 2x}{\cos x} = \sec x.$

17. $\dfrac{\tan(x + y)}{\cot(x - y)} = \dfrac{\tan^2 x - \tan^2 y}{1 - \tan^2 x \tan^2 y}.$

18. $\tan(x + y + z) = \dfrac{\tan x + \tan y + \tan z - \tan x \tan y \tan z}{1 - \tan x \tan y - \tan y \tan z - \tan x \tan z}.$

19. Derive the following formulas:

   (a) $\sin u \cos v = \frac{1}{2}[\sin(u + v) + \sin(u - v)]$

   (b) $\sin u \sin v = \frac{1}{2}[\cos(u - v) - \cos(u + v)]$

   (c) $\cos u \cos v = \frac{1}{2}[\cos(u + v) + \cos(u - v)].$

20. Make appropriate substitutions in the formulas of the previous problem to obtain:

   (a) $\sin x + \sin y = 2\sin\left(\dfrac{x + y}{2}\right)\cos\left(\dfrac{x - y}{2}\right)$

   (b) $\cos y - \cos x = 2\sin\left(\dfrac{x + y}{2}\right)\sin\left(\dfrac{x - y}{2}\right)$

(c) $\cos x + \cos y = 2 \cos \left(\dfrac{x+y}{2}\right) \cos \left(\dfrac{x-y}{2}\right)$.

21. $\dfrac{\sin x + \sin 2x}{\cos x + \cos 2x} = \tan \tfrac{3}{2}x$.

15. Even and Odd Functions

The properties possessed by the sine and cosine functions as given in Equations (14.3) and (14.4), $\sin(-t) = -\sin t$ and $\cos(-t) = \cos t$, occur elsewhere in mathematics, and are important enough to warrant a discussion. We have the following definition.

Definition 15.1. Let f be a function such that $f(-x)$ is defined whenever $f(x)$ is. f is called an *even* function, if and only if $f(-x) = f(x)$ for all x in the domain of f. f is called an *odd* function, if and only if $f(-x) = -f(x)$ for all x in the domain of f.

From this definition, it follows that the sine function is an odd function, and the cosine function is an even function. Many other examples exist, a few of which we list below.

Example 15.2. The identity function I is an odd function, because $I(x) = x$, hence $I(-x) = -x = -I(x)$.

Example 15.3. The function g defined by $g(x) = x^{2k}$, where k is a positive integer, is even. For

$$g(-x) = (-x)^{2k} = [(-x)^2]^k = (x^2)^k = x^{2k} = g(x).$$

Example 15.4. The function f defined by $f(x) = x + 1$ is neither even nor odd, for $f(-x) = -x + 1$. Thus, for example, $f(-1) = 0$, while $f(1) = 2, f(-2) = -2 + 1 = -1$ and $f(2) = 2 + 1 = 3$, and so on.

While this example shows that a function need not be either even or odd, the following theorem relates all functions with suitable domains to even and odd functions.

Theorem 15.5. Let f be any function satisfying the condition that if $f(x)$ is defined, then so is $f(-x)$. Then for each x in the domain of f,

$$f(x) = u(x) + v(x),$$

where u is an even function and v is an odd function.

Proof. For each x in the domain of f, consider the identity

$$f(x) = \tfrac{1}{2}[f(x) + f(-x)] + \tfrac{1}{2}[f(x) - f(-x)].$$

Then let u be defined by $u(x) = \tfrac{1}{2}[f(x) + f(-x)]$ and v be defined by $v(x) = \tfrac{1}{2}[f(x) - f(-x)]$. Direct substitution gives

$$u(-x) = \tfrac{1}{2}[f(-x) + f(x)] = u(x)$$

and $$v(-x) = \tfrac{1}{2}[f(-x) - f(x)] = -v(x),$$

which proves that u is even and v is odd.

From the proof of the theorem, we also obtain a method for constructing the functions in the conclusion of the theorem.

We list below other theorems on even and odd functions. The proofs are straightforward, and most of them will be left as exercises. As the zero function satisfies the conditions defining both even and odd functions, it will be assumed that f and g are not identically zero.

Theorem 15.6. If f and g are both even (odd) functions, then:

(i) Their sum or difference $f \pm g$ defined by $(f \pm g)(x) = f(x) \pm g(x)$ is even (odd).

(ii) Their product fg defined by $(fg)(x) = f(x)g(x)$ is even.

(iii) Their quotients defined by

$$(f/g)(x) = \frac{f(x)}{g(x)}$$

and $$(g/f)(x) = \frac{g(x)}{f(x)}$$

are both even.

(iv) The composition $f \circ g$ defined by $(f \circ g)(x) = f[g(x)]$ is even (odd).

Theorem 15.7. If f is an even function and g is an odd function, then:

(i) Their sum or difference, $f \pm g$, is neither even nor odd.

(ii) Their product fg is odd.

(iii) The quotients f/g and g/f are both odd.

(iv) The compositions $f \circ g$ and $g \circ f$ are both even.

Proof of Theorem 15.6, part (iv): As $(f \circ g)(x) = f[g(x)]$, if f and g are both even, then for each x,

$$(f \circ g)(-x) = f[g(-x)]$$
$$= f[g(x)]$$
$$= (f \circ g)(x).$$

If both f and g are odd, on the other hand,

$$(f \circ g)(-x) = f[g(-x)]$$
$$= f[-g(x)]$$
$$= -f[g(x)]$$
$$= -(f \circ g)(x).$$

EXERCISES

1. Determine whether the following equations define functions which are even, odd, or neither even nor odd. Justify your answers.
   (a) $f(x) = x^2 + 2x - 1$
   (b) $g(x) = x^4(x^3 + x)$
   (c) $f(x) = \sin x \cos x$
   (d) $f(x) = |x|$
   (e) $f(x) = x - [x]$
   (f) $f(x) = \sin^2 x + \sec x$
   (g) $f(x) = 2^{\sin x}$
   (h) $f(x) = 2^{\cos x}$
   (i) $f(x) = \log(2 + \cos x)$.

2. Prove Theorem 15.6, (i)–(iii).

3. Prove Theorem 15.7, (i)–(iv).

4. Which of the trigonometric functions are even? odd?

5. (a) If f is an even function and the point (x,y) is in the graph of f, then which of the following must be in the graph of f, (y,x), $(-x,y)$, $(-x,-y)$, $(-y,-x)$, $(-y,x)$? (b) The same if f is odd.

6. The graph of a function f is said to be *symmetric* with respect to: (a) the y axis if $(-x,y)$ is in the graph of f whenever (x,y) is; (b) the origin if $(-x,-y)$ is in the graph of f whenever (x,y) is. What is the relation between even and odd functions and these two symmetry concepts?

7. Use the ideas of Problem 6 to aid in sketching the graphs of the functions f and g, where $f(x) = 2^x - 2^{-x}$ and $g(x) = x^2 + 1$.

16. Formulas at Multiples and Fractions of *t*

The following formulas are immediate in view of the formulas of Section 14 and are left as exercises for the reader.

$$(16.1) \qquad \sin 2t = 2 \sin t \cos t,$$

$$(16.2) \qquad \cos 2t = \cos^2 t - \sin^2 t,$$

$$(16.3) \qquad \qquad = 1 - 2 \sin^2 t,$$

$$(16.4 \qquad \qquad = 2 \cos^2 t - 1,$$

$$(16.5) \qquad \tan 2t = \frac{2 \tan t}{1 - \tan^2 t}.$$

From these formulas we are able to complete some of the previous sections. For example, from Equation (16.1) it follows that the product function of the sine and cosine functions has primitive period $2\pi/2 = \pi$. Since

$$\cos^2 t = \frac{1 + \cos 2t}{2},$$

the function f defined by $f(t) = \cos^2 t$ is also periodic with primitive period π. [Note that this is an application of Theorem 11.2, where $g(x) = x^2$, $f(t) = \cos t$ and $h(x) = g[f(x)]$. In that theorem we showed that h had primitive period $\leq$ the primitive period of f, but had no example to illustrate that h could have primitive $<$ the primitive period of f.]

Using formulas (16.3) and (16.4) above we are able to derive formulas for $\sin (t/2)$, $\cos (t/2)$, and $\tan (t/2)$. We first obtain $\sin (t/2)$ and $\cos (t/2)$.

First of all

$$2 \cos^2 t = \cos 2t + 1,$$

or

$$\cos^2 t = \frac{\cos 2t + 1}{2}.$$

Making the substitution $t = u/2$, gives

$$\cos^2 \frac{u}{2} = \frac{1 + \cos u}{2}.$$

Then extracting square roots from both sides, we write

$$\left|\cos\frac{u}{2}\right| = \sqrt{\frac{1 + \cos u}{2}}.$$

Thus,

(16.6)
$$\cos\frac{u}{2} = (\text{sign})\sqrt{\frac{1 + \cos u}{2}},$$

where (sign) indicates that the radical is to be preceded by the appropriate algebraic sign, depending, of course, on the particular number $u/2$.

From the formula

$$\cos 2t = 1 - 2\sin^2 t$$

we have

$$2\sin^2 t = 1 - \cos 2t,$$

or
$$\sin^2 t = \frac{1 - \cos 2t}{2}.$$

Then, letting $t = u/2$ yields

$$\sin^2\frac{u}{2} = \frac{1 - \cos u}{2}.$$

Extracting square roots from both sides of this equation gives us

$$\left|\sin\frac{u}{2}\right| = \sqrt{\frac{1 - \cos u}{2}}.$$

Thus,

(16.7)
$$\sin\frac{u}{2} = (\text{sign})\sqrt{\frac{1 - \cos u}{2}},$$

where (sign) is to be interpreted as above.

From the relation $\tan t = \sin t/\cos t$, and Equations (16.6) and (16.7) above we have

(16.8)
$$\tan\frac{u}{2} = \frac{\sin\dfrac{u}{2}}{\cos\dfrac{u}{2}} = (\text{sign})\sqrt{\frac{1 - \cos u}{1 + \cos u}},$$

the appropriate sign depending on $u/2$. The denominator of the radicand in Equation (16.8) can be rationalized in two different ways. One is obtained by:

$$\sqrt{\frac{1 - \cos u}{1 + \cos u}} = \sqrt{\frac{(1 - \cos u)(1 + \cos u)}{(1 + \cos u)^2}}$$

$$= \sqrt{\frac{1 - \cos^2 u}{(1 + \cos u)^2}}$$

$$= \frac{\sqrt{\sin^2 u}}{1 + \cos u}$$

$$= \frac{|\sin u|}{1 + \cos u}.$$

It is readily verified that $\tan (u/2)$ and $\sin u$ have the same sign. Hence,

(16.9) $$\tan \frac{u}{2} = \frac{\sin u}{1 + \cos u}.$$

As

$$\frac{\sin u}{1 + \cos u} = \frac{1 - \cos u}{\sin u}$$

is an identity, it follows that

(16.10) $$\tan \frac{u}{2} = \frac{1 - \cos u}{\sin u}.$$

We now consider some examples applying the results of this section.

Example 16.1. Evaluate $\sin (5\pi/12)$.

Solution. $5\pi/12 = \frac{1}{2}(5\pi/6)$. Thus from Equation (16.7),

$$\sin \frac{5\pi}{12} = \sqrt{\frac{1 - \cos \frac{5\pi}{6}}{2}}$$

$$= \sqrt{\frac{1 - \left(-\frac{\sqrt{3}}{2}\right)}{2}}$$

$$= \sqrt{\frac{1 + \frac{\sqrt{3}}{2}}{2}}$$

$$= \sqrt{\frac{2 + \sqrt{3}}{4}}$$

$$= \frac{1}{2}\sqrt{2 + \sqrt{3}}.$$

Example 16.2. Given that tan $t = \frac{5}{4}$ and sin $t < 0$, find tan $(t/2)$.

Solution. As tan $t = \frac{5}{4}$ and sin $t < 0$, we may write sin $t = -5k$ and cos $t = -4k$, where $k > 0$. Then as $\sin^2 t + \cos^2 t = 1$, we have $25k^2 + 16k^2 = 1$, or $k^2 = \frac{1}{41}$. Thus, $k = 1/\sqrt{41}$, and sin $t = -5/\sqrt{41}$, cos $t = -4/\sqrt{41}$. Then from formula (16.10),

$$\tan \frac{t}{2} = \frac{1 - \cos t}{\sin t}$$

$$= \frac{1 - \left(-\dfrac{4}{\sqrt{41}}\right)}{-\dfrac{5}{\sqrt{41}}}$$

$$= \frac{\sqrt{41} + 4}{-5}$$

$$= -\frac{\sqrt{41} + 4}{5}.$$

EXERCISES

1. Derive Equation (16.10) directly from (16.8).

2. Find cos $\pi/12$ by two different methods. Verify that the answers are equal.

3. Supply the details necessary to derive Equations (16.1)–(16.8).

4. Evaluate each of the following:

(a) $\sin \dfrac{3\pi}{8}$ (c) $\cos \dfrac{\pi}{8}$ (e) $\tan \dfrac{\pi}{12}$

(b) $\sin \dfrac{\pi}{8}$ (d) $\cos \dfrac{\pi}{16}$ (f) $\tan \dfrac{5\pi}{12}$.

Prove the following identities.

5. $\sin 3t = 3 \sin t - 4 \sin^3 t$.

6. $\cos 3t = 4 \cos^3 t - 3 \cos t$.

7. $\cot t - \tan t = 2 \cot 2t$.

8. $\cos^4 t - \sin^4 t = \cos 2t$.

9. $\cot (x + y) = \dfrac{\cot x \cot y - 1}{\cot y + \cot x}$.

10. $\cot (x - y) = \dfrac{\cot x \cot y + 1}{\cot y - \cot x}$.

11. $\cos x = \dfrac{1 - \tan^2 \dfrac{x}{2}}{1 + \tan^2 \dfrac{x}{2}}.$

12. $2 \tan 2x = \dfrac{\cos x + \sin x}{\cos x - \sin x} - \dfrac{\cos x - \sin x}{\cos x + \sin x}.$

13. $\sin^4 x = \frac{3}{8} - \frac{1}{2} \cos 2x + \frac{1}{8} \cos 4x.$

14. $\sin(x + y) - \sin(x - y) = 2 \cos x \sin y.$

15. $\tan x \sin 2x = 2 \sin^2 x.$

16. $\cot x \sin 2x = 1 + \cos 2x.$

17. $\dfrac{1 - \sin 2x}{\cos 2x} = \dfrac{1 - \tan x}{1 + \tan x}.$

18. $\cos 2x = \dfrac{1 - \tan^2 x}{1 + \tan^2 x}.$

19. $\cos^4 x = \frac{3}{8} + \frac{1}{2} \cos 2x + \frac{1}{8} \cos 4x.$

20. $\tan \frac{1}{2}x = \csc x - \cot x.$

21. $\dfrac{2 \tan \frac{1}{2}x}{1 + \tan^2 \frac{1}{2}x} = \sin x.$

22. $\dfrac{1 + \cos 2x}{\sin 2x} = \cot x.$

23. $\dfrac{\sin^3 t - \cos^3 t}{\sin t - \cos t} = 1 + \frac{1}{2} \sin 2t.$

24. $\dfrac{1 - \tan \frac{1}{2}x}{1 + \tan \frac{1}{2}x} = \dfrac{1 - \sin x}{\cos x}.$

25. $\dfrac{\sin x + \cos 2x - 1}{\cos x - \sin 2x} = \tan x.$

26. $\dfrac{\sin 4x}{\sin 2x} = 2 \cos 2x.$

27. $\dfrac{2 \cot x}{\csc^2 x - 2} = \tan 2x.$

28. $\dfrac{1 - \cos x - \tan^2 \frac{1}{2}x}{\sin^2 \frac{1}{2}x} = \dfrac{2 \cos x}{1 + \cos x}.$

29. $\dfrac{\sin 5x}{\sin x} - \dfrac{\cos 5x}{\cos x} = 4 \cos 2x.$

3

GRAPHS OF THE TRIGONOMETRIC FUNCTIONS

17. Least Upper Bounds and Greatest Lower Bounds of Functions

The concepts of least upper bounds and greatest lower bounds are extremely important in mathematics. Although the student at this level will undoubtedly not understand all the details, we feel it is essential to give the formal definitions together with some intuitive devices. We first consider some examples to illustrate the concepts before defining them.

Example 17.1. Let f be defined by $f(x) = x$, where $0 \leq x \leq 1$, i.e., $f = \{(x,y) : y = x, 0 \leq x \leq 1\}$. The function f has the property that there exist constants m and M such that for all x in the domain of f, $m \leq f(x) \leq M$. For example, $f(x) \leq 2$ and $f(x) \geq -2$ for $0 \leq x \leq 1$. Here there is a smallest M, such that $f(x) \leq M$, namely $M = 1$, and a largest m such that $m \leq f(x)$, namely $m = 0$. Observe that $f(0) = 0$

and $f(1) = 1$, so it follows that $f(0) \le f(x) \le f(1)$ for all x in the domain of f.

Example 17.2. Let $g = \{(x,y) : y = x, 0 < x < 1\}$. As in Example 17.1, there are constants M and m such that for all x in the domain of g, $m \le g(x) \le M$. The smallest such M is $M = 1$, and the largest such m is $m = 0$, as above. In particular, then, $0 < g(x) < 1$ for all x in the domain of g, while both $g(x) \ne 1$ and $g(x) \ne 0$ for all x.

Example 17.3. Let $h = \{(x,y) : y = 1/x, x > 0\}$. Then for every x in the domain of h, $h(x) > 0$. Observe that $1/x$ can be made as large as we please by making x sufficiently small. Thus, there is no constant M such that $h(x) \le M$ for all x in the domain of h. In the same manner, $1/x$ can be made an arbitrarily small positive number by choosing x sufficiently large. Hence, $h(x) > 0$ and 0 is the largest number m such that $h(x) > m$ for all x in the domain of h. $h(x) \ne 0$ for all x in the domain of h, however.

Observe that just the reverse of the results of Example 17.3 can be obtained by letting $k = \{(x,y) : y = -(1/x), x > 0\}$. Then $k(x) = -h(x)$, and $k(x) < 0$ for all x; and $k(x)$ can be made as small as desired by taking x to be a sufficiently small positive number.

We now give formal definitions to the concepts encountered in these examples.

Definition 17.4. A function f is said to be *bounded above*, or to have an *upper bound*, if and only if there is a constant M such that $f(x) \le M$ for all x in the domain of f. f is *bounded below*, or has a *lower bound*, if and only if there is a constant m such that $m \le f(x)$ for all x in the domain of f. f is said to be *bounded*, if and only if it is bounded above and below.

Definition 17.5. The number M is called the *least upper bound* of the function f, abbreviated l.u.b. f, if and only if M is an upper bound of f, and no number less than M is an upper bound of f. The number m is called the *greatest lower bound* of the function f, abbreviated g.l.b. f, if and only if m is a lower bound of f, and no number greater than m is a lower bound. If $M = $ l.u.b. f and $f(x) = M$ for some x, then M is called the *maximum* of f. Similarly, if $m = $ g.l.b. f and $f(x) = m$ for some x, then m is called the *minimum* of f.

Note. If f has no upper bound, then l.u.b. f does not exist. Similarly g.l.b. f does not exist, when f is not bounded below. However, if f has a lower bound, it has a greatest lower bound; if it has an upper bound, it has a least upper bound.

In Example 17.1, f has both a maximum and a minimum. In Example 17.2, g has neither a maximum nor a minimum. However, g.l.b. g = 0 and l.u.b. g = 1, so g is bounded. In Example 17.3, h is unbounded, as it has no upper bound. h has a lower bound, however, and g.l.b. h = 0. h has no minimum, however, because $h(x) > 0$ for all x.

The two definitions given here have some geometrical significance. For if m = g.l.b. f, then the graph of f does not go below the line $y = m$. Similarly, the graph of f does not go above the line y = l.u.b. f. Thus, l.u.b. f and g.l.b. f are useful in sketching the graph of the function f. For periodic functions we can go even further than the above definitions.

Definition 17.6. Let f be a bounded periodic function with l.u.b. f = M and g.l.b. $f = m$. Then the *amplitude* of f is defined by

$$\text{amplitude } f = \tfrac{1}{2}(M - m).$$

In case f is an unbounded periodic function, then f is said to have *infinite* amplitude.

Example 17.7. The sine function has amplitude 1, for l.u.b. sine = 1 and g.l.b. sine = -1. Thus amplitude sine = $\tfrac{1}{2}[1 - (-1)] = \tfrac{1}{2}(2) = 1$.

Example 17.8. The tangent function has infinite amplitude, for the tangent is unbounded.

EXERCISES

Find the l.u.b. and g.l.b. (if they exist) of the functions defined by the following. In each case state whether the l.u.b. and g.l.b. are maximum and minimum.

1. cosine

2. $g(x) = \sec x, 0 \leq x < \dfrac{\pi}{2}.$

3. $f(x) = x^2, -1 \leq x \leq 2.$

4. $g(x) = x^3, -1 \leq x < 1.$

5. $f(x) = 1 + \sin x.$

6. $g(x) = 1 - \cos x.$

7. $h(x) = \sin^2 x.$

8. $F(x) = 2^{\sin x}$.

9. $G(x) = \cot x, 0 < x \le \dfrac{\pi}{2}.$

Find the period and amplitude of the functions defined by the following:

10. $f(x) = \sin 3x$.

11. $g(x) = \sec x/2$.

12. $h(x) = \begin{cases} 2, 0 < x < 1 \\ -1, 1 < x < 2. \end{cases}$

$\qquad h(x + 2) = h(x)$ for all real x.

13. $f(x) = |x|, -1 \le x \le 1, f(x + 2) = f(x)$ for all x.

18. The Graph of y = sin x

As the graph of the function f defined by the equation $y = f(x)$ is the same as the graph of the equation $y = f(x)$, these two concepts will be used interchangeably. Thus, the graph of the sine function is the same as the graph of the equation $y = \sin x$. The graph will be obtained from the unit circle in the manner below and its construction explained briefly.

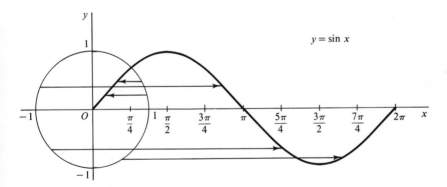

Figure 18.1

The unit circle is placed in its usual position with center at the origin. Now for each x, $0 \le x \le 2\pi$, the ordinate on the graph of the sine function is precisely the ordinate at $P(t)$ for $t = x$. Thus, to obtain the graph of $y = \sin x$, extend a horizontal segment from $P(x)$ to the point whose abscissa is x. To obtain the entire graph of the function, it is only necessary to observe that the sine function has primitive period 2π. Hence, the graph between 2π and 4π, for example, is precisely the same as between 0 and 2π.

19. The Graph of $y = A \sin (ax + b)$

Beginning with the graph of $y = \sin x$, we obtain the graph of $y = \sin ax$ in the following manner. If f is defined by $f(x) = \sin x$ and g is defined by $g(x) = f(ax) = \sin ax$, then recall from section **11**, Theorem 11.3, that g has primitive period $2\pi/a$. As the sine function is an odd function, we need only obtain the graph of $y = \sin ax$ through one-half the period, namely, from 0 to π/a. For if $-\pi/a \le x \le 0$, then $\sin ax = -\sin a(-x)$ for $0 \le -x \le \pi/a$. The following relations enable us to immediately sketch the graph for $0 \le x \le \pi/a$:

$$\text{If} \begin{cases} \text{(i)} \quad 0 \le x \le \dfrac{\pi}{2a}, \quad \text{then} \quad 0 \le ax \le \dfrac{\pi}{2}. \\[2mm] \text{(ii)} \quad \dfrac{\pi}{2a} \le x \le \dfrac{\pi}{a}, \quad \text{then} \quad \dfrac{\pi}{2} \le ax \le \pi. \end{cases}$$

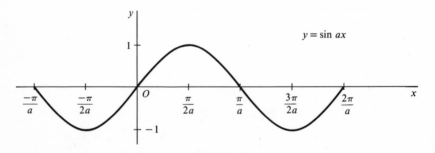

Figure 19.1

Thus, the graph in general is given by Figure 19.1.

Note. The graph of $y = \sin t$ with $t = ax$ is precisely the graph of $y = \sin ax$.

Example 19.1. Sketch the graphs of $y = \sin 2x$, $y = \sin x$, and $y = \sin x/2$ on the same coordinate axes.

As the function f defined by $f(x) = \sin 2x$ has primitive period $2\pi/2 = \pi$, the graph of $y = \sin 2x$ is completely determined by sketching it between 0 and π. On the other hand the function g defined by $g(x) = \sin x/2$

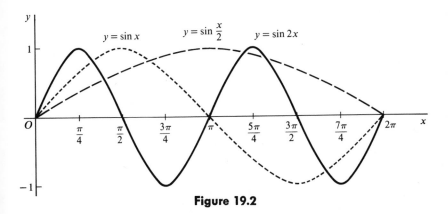

Figure 19.2

has primitive period 4π, so only half its graph is obtained on the interval between 0 and 2π. An examination of Figure 19.2 shows precisely what happens in general. The graph of $y = \sin ax$ is obtained by "squeezing" the graph of $y = \sin x$ when $a > 1$ and by "stretching" it when $a < 1$.

The graph of $y = A \sin ax$ is easily obtained from the graph of $y = \sin ax$. For if (x,y) is in the graph of $y = \sin ax$, then (x,Ay) is in the graph of $y = A \sin ax$. Figure 19.3 below illustrates this for $A > 1$.

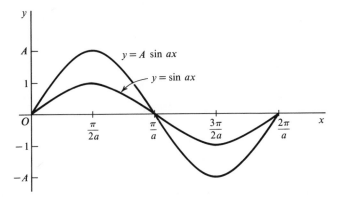

Figure 19.3

To complete the section, we wish to sketch the graph of $y = A \sin (ax + b)$. From the above remarks, we need only examine the special case $A = 1$.

The first thing to do is find the primitive period of the function f defined by $f(x) = \sin (ax + b)$. Clearly $\sin (ax + b + 2\pi) = \sin (ax + b)$ for all x, so f is periodic. If $g(x) = \sin ax$, g has primitive period $2\pi/a$, as previously observed. Suppose the primitive period of f is p. Then

$$f(x + p) = \sin [a(x + p) + b] = \sin (ax + ap + b)$$

$$= \sin (ax + b) = f(x).$$

It follows that $ax + ap + b = ax + 2\pi + b$, and hence, $ap = 2\pi$ or $p = 2\pi/a$. Thus, the primitive period of f is $2\pi/a$.

To sketch the graph of $y = \sin (ax + b)$, we proceed as follows. The graph of $y = \sin at$ has its starting point at $(0,0)$, i.e., when $t = 0$, $y = 0$. Thus , we consider the real number $t = x + b/a$ and the relations:

(i)   If $0 \le t \le \dfrac{\pi}{2a}$,   then $0 \le x + \dfrac{b}{a} \le \dfrac{\pi}{2a}$

and $\dfrac{-b}{a} \le x \le \dfrac{\pi}{2a} - \dfrac{b}{a}$.

(ii) If $\dfrac{\pi}{2a} \le t \le \dfrac{\pi}{a}$,   then $\dfrac{\pi}{2a} \le x + \dfrac{b}{a} \le \dfrac{\pi}{a}$

and $\dfrac{\pi}{2a} - \dfrac{b}{a} \le x \le \dfrac{\pi}{a} - \dfrac{b}{a}$.

(iii) If $\dfrac{\pi}{a} \le t \le \dfrac{3\pi}{2a}$,   then $\dfrac{\pi}{a} \le x + \dfrac{b}{a} \le \dfrac{3\pi}{2a}$

and $\dfrac{\pi}{a} - \dfrac{b}{a} \le x \le \dfrac{2\pi}{2a} - \dfrac{b}{a}$.

(iv) If $\dfrac{3\pi}{2a} \le t \le \dfrac{2\pi}{a}$,   then $\dfrac{3\pi}{2a} \le x + \dfrac{b}{a} \le \dfrac{2\pi}{a}$

and $\dfrac{3\pi}{2a} - \dfrac{b}{a} \le x \le \dfrac{2\pi}{a} - \dfrac{b}{a}$.

From this set of relations it is clear that the graph of $y = \sin (ax + b)$ is merely the graph of $y = \sin ax$ moved $|b/a|$ units to right or left, depending on the sign of b. If $b > 0$ the graph is shifted to the left, and if $b < 0$ the graph is shifted to the right. It is for these reasons that $|b/a|$ is called the *phase shift* of the graph.

To summarize briefly, there are three concepts which are very useful in sketching the graph of $y = A \sin (ax + b)$. They are the (primitive) period, $2\pi/a$, the amplitude $|A|$, and the phase shift, $|b/a|$. A fourth concept, called the frequency, is defined to be the reciprocal of the period, and gives, intuitively, the fractional part of the graph that appears in an interval of unit length. We will now apply these to an example.

Example 19.1.   Sketch the graph of $y = 2 \sin (3x - \pi/4)$.

Solution.   The period is $2\pi/3$, the amplitude is 2, the phase shift is $\pi/12$ to the right, and the frequency is $3/2\pi$. Thus the graph is that of $y = 2 \sin 3x$ moved $\pi/12$ units to the right of the origin.

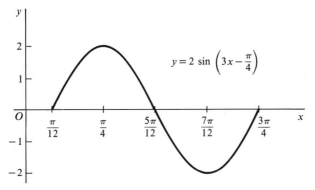

Figure 19.4

A final word may aid the student in sketching the graph of $y = A \sin (ax + b)$. Set $ax + b = 0$. Then $x = -b/a$ is the starting point of the graph. As the period is $2\pi/a$, lay off a segment of this length to the right of $-b/a$, thus giving an end point of $(2\pi/a) - (b/a)$. If this segment is divided into four equal parts, by the points

$$\frac{\pi}{2a} - \frac{b}{a}, \frac{\pi}{a} - \frac{b}{a} \quad \text{and} \quad \frac{3\pi}{2a} - \frac{b}{a},$$

the graph crosses the x axis at $(\pi/a) - (b/a)$ and reaches its maximum and minimum points at

$$\frac{\pi}{2a} - \frac{b}{a} \quad \text{and} \quad \frac{3\pi}{2a} - \frac{b}{a},$$

respectively.

EXERCISES

Find the period, amplitude, and phase shift of each of the following. Sketch the graph through an interval of length equal to the period.

1. $y = 2 \sin 3x$.

2. $y = 2 \sin (3x + 1)$.

3. $y = \frac{1}{2} \sin \pi x$.

4. $y = 3 \sin (\pi x - 2)$.

5. $y = \sin \left(\dfrac{x}{2} - \dfrac{\pi}{2} \right)$.

6. $y = -\sin (x - 1)$.

7. $y = \frac{3}{4} \sin (2x + 3)$.

8. $y = -2 \sin (2\pi x - \pi)$.

9. $y = 4 \sin (2x - 1)$.

10. $y = 2 \sin \left(2x - \dfrac{\pi}{3} \right)$.

11. $y = -3 \sin (\pi x + \frac{1}{8})$.

12. $y = -2 \sin \left(3x + \dfrac{\pi}{2} \right)$.

13. $y = \sin \left(\dfrac{1}{\pi} x + \frac{1}{2} \right)$.

14. $y = -\sin \left(\dfrac{\pi}{4} x - 2 \right)$.

15. $y = 5 \sin \left(\frac{3}{4} x + \dfrac{\pi}{8} \right)$.

16. $y = -2 \sin \left(\frac{2}{3} x - \dfrac{\pi}{9} \right)$.

20. The Graphs of $y = \cos x$, $y = \tan x$, and $y = \sec x$

Using the results of section **19**, it is possible to sketch immediately the graph of $y = \cos x$. For $\sin (x + \pi/2) = \cos x$, and the graph of $y = \sin (x + \pi/2)$ is the graph of $y = \sin x$ shifted $\pi/2$ units to the left of the origin. The graph of $y = \cos x$ is sketched below for $-\pi/2 \le x \le 2\pi$. The rest of the graph is obtained using the fact that the cosine function has period 2π. A sketch of the graph of $y = A \cos (ax + b)$ can be obtained from the graph of $y = \cos x$ in a manner similar to the method applied in section **19**.

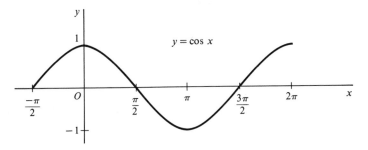

Figure 20.1

To sketch the graph of $y = \tan x$, we will use the graphs of $y = \sin x$ and $y = \cos x$. As $\tan x = \sin x/\cos x$, we obtain the ordinate for each x as the quotient $\sin x/\cos x$. The tangent function is an odd function and has primitive period π. These two facts enable us to determine the graph in its entirety from the graph between 0 and $\pi/2$. (See Figures 20.2 and 20.3.)

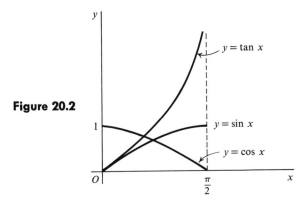

Figure 20.2

To sketch the graph of $y = \sec x$, the fact that $\sec x = 1/\cos x$ is used together with the graph of $y = \cos x$. Observe that if $(x,y) \in$ cosine, i.e., $y = \cos x$, then $(x,1/y) \in$ secant or $1/y = \sec x = 1/\cos x$. Thus, the ordinate of the point on the graph of $y = \sec x$ is the reciprocal of the ordinate of the corresponding point on the graph of $y = \cos x$. The graph is given in Figure 20.4.

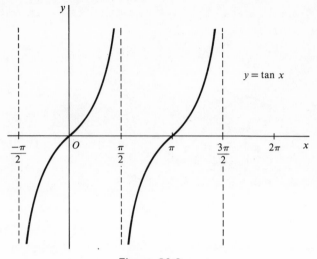

Figure 20.3

The vertical broken lines through $\pi/2$, $3\pi/2$, etc., in the graphs of the secant and tangent functions serve a useful purpose. They are called *vertical asymptotes*, and the graphs do not cross these lines while getting "arbitrarily close" to them. The student should also observe that the secant function is unbounded, and in fact $|\sec x| \geq 1$ for all x for which $\sec x$ is defined.

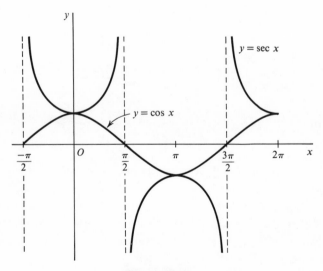

Figure 20.4

EXERCISES

Sketch the graphs of the following, giving first the period, amplitude, and phase shift.

1. $y = 2 \cos 2x$.

2. $y = 2 \cos (2x - 1)$.

3. $y = 3 \cos \left(\dfrac{x}{2} + 1 \right)$.

4. $y = 2 \sec x$.

5. $y = 2 \sec \left(x - \dfrac{\pi}{4} \right)$.

6. $y = 3 \sec 2x$.

7. $y = \frac{1}{4} \sec 2x$.

8. $y = 4 \cos (\pi x - 2)$.

9. $y = \tan 2x$.

10. $y = 2 \tan (3x - 1)$.

11. $y = \frac{1}{2} \tan \dfrac{x}{2}$.

12. $y = \tan (\pi x + 2)$.

13. Sketch the graph of $y = \csc x$ for $-\dfrac{\pi}{2} \le x \le \dfrac{3\pi}{2}$, $x \ne 0, \pi$, using the graph of $y = \sin x$.

14. Sketch the graph of $y = \cot x$ for $0 < x < \pi$, using a method similar to one of those applied in this section.

21. Special Graphing Methods: Composition of Ordinates and Use of Identities

These two methods will be illustrated below by examples, the first of which uses composition of ordinates.

Example 21.1. Sketch the graph of the equation $y = x + \sin x$.

Solution.   To obtain the graph of this equation, we consider two equations, $y_1 = x$ and $y_2 = \sin x$, and sketch them on the same coordinate axes. Then if (x,y_1) and (x,y_2) are points in the graphs of the respective equations, $(x,y_1 + y_2)$ is in the graph of $y = x + \sin x$. Briefly, then, the graph of $y = x + \sin x$ is obtained by adding the ordinates of the separate graphs corresponding to the same abscissa, hence the name "composition of ordinates." A sketch of the graph is given below. The broken lines are the graphs of y_1 and y_2, and the solid curve is that of their sum.

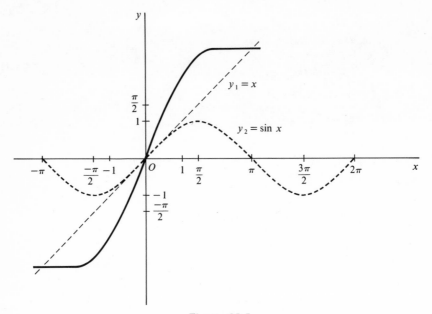

Figure 21.1

The method of composition of ordinates is useful, not just in sketching graphs involving trigonometric functions, but in other areas as well. One drawback in using the composition of ordinates is that, in many instances, a large number of points must be plotted to give a fairly accurate sketch of the graph. Nevertheless, in some cases, it is the only method we have at our disposal.

The two examples that follow illustrate two uses of identities in curve sketching.

Example 21.2. Sketch the graph of the equation $y = \sin x \cos x$.

Solution. As $\sin 2x = 2 \sin x \cos x$, the graph of $y = \sin x \cos x$ is the same as the graph of $y = \frac{1}{2} \sin 2x$. This is of the form $y = A \sin ax$ with $A = \frac{1}{2}$ and $a = 2$. The curve is sketched below for $0 \leq x \leq 2\pi$.

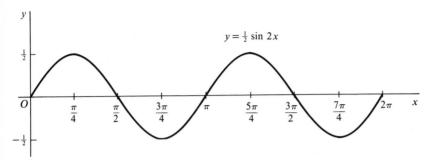

Figure 21.2

Example 21.3. To sketch the graph of the equation $y = A \sin ax + B \cos ax$, observe that

$$A \sin ax + B \cos ax =$$
$$\sqrt{A^2 + B^2} \left[\frac{A}{\sqrt{A^2 + B^2}} \sin ax + \frac{B}{\sqrt{A^2 + B^2}} \cos ax \right].$$

Now
$$\left[\frac{A}{\sqrt{A^2 + B^2}} \right]^2 + \left[\frac{B}{\sqrt{A^2 + B^2}} \right]^2 = 1,$$

and hence there is a real number b, $0 \le b < 2\pi$, such that

$$\cos b = \frac{A}{\sqrt{A^2 + B^2}}$$

and
$$\sin b = \frac{B}{\sqrt{A^2 + B^2}}.$$

Then we may write

$$y = A \sin ax + B \cos ax$$
$$= \sqrt{A^2 + B^2} \,(\sin ax \cos b + \cos ax \sin b)$$
$$= \sqrt{A^2 + B^2} \sin (ax + b).$$

Thus, the problem has been reduced to sketching the graph of an equation of the form $y = C \sin (cx + d)$. Here the period is $2\pi/a$, the amplitude is $\sqrt{A^2 + B^2}$, and the phase shift is b/a units to the left using the convention of section **19**.

A specific example using this last method seems advisable, so let us sketch $y = \sin x + 2 \cos x$. Then $A = 1$, $B = 2$, $A^2 + B^2 = 5$. Thus,

$$y = \sqrt{5}\left[\frac{1}{\sqrt{5}}\sin x + \frac{2}{\sqrt{5}}\cos x\right].$$
$$= \sqrt{5}\,(\sin x \cos b + \cos x \sin b)$$
$$= \sqrt{5}\,\sin (x + b).$$

If $\cos b = 1/\sqrt{5} = \sqrt{5}/5$ or $\cos b$ is approximately 0.447 and $\sin b > 0$, then b is approximately 1.11 from Table II (page 160). Thus, $y = \sin x + 2\cos x = \sqrt{5}\sin (x + 1.11)$. The period is 2π, the amplitude is $\sqrt{5}$, and the phase shift is 1.11 to the left. The graph is sketched below on the interval $-1.11 \le x \le 2\pi - 1.11$.

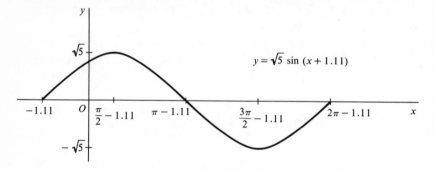

Figure 21.3

EXERCISES

Find the period, amplitude, phase shift and sketch the graphs of the following equations.

1. $y = \sin x + \cos x.$

2. $y = \sin 2x + \sqrt{3}\cos 2x.$

3. $y = \cos^2 x.$

4. $y = \sin^2 x.$

5. $y = 1 + \sin x - \cos\left(\frac{\pi}{2} - x\right).$

Sketch the following using composition of ordinates.

6. $y = x - \cos x.$

7. $y = \sin x + \sin 2x.$

8. $y = \cos x + \sin 2x.$

9. $y = 1 + \sin x.$

10. $y = 1 - \cos x.$

4

ANGLES AND APPLICATIONS

22. Angles

This section gives a brief discussion of angles and definitions which provide meaning to such expressions as "sin α," where α is an angle.

First of all, an angle is formed in the plane when a ray is rotated about its end point. The ray in its original position is called the *initial side;* in its final position, it is called the *terminal side;* and the end point is called the *vertex* of the angle. An angle is said to be in *standard position* if its initial side is on the positive x axis with its vertex at the origin.

Figure 22.1 illustrates an angle in standard position with the angle θ, or its terminal side, in the first quadrant.

Let $C = \{(x,y) : x^2 + y^2 = 1\}$ as before, and let θ be an angle in standard position. As the angle θ is formed by rotating the positive x axis about the origin, the unit point moves along C. The measure of θ in radians is plus or minus the arc length, through which the unit point moves in the generation of θ. The sign is plus, if the rotation is counter-clockwise, and minus, if the rotation is clockwise.

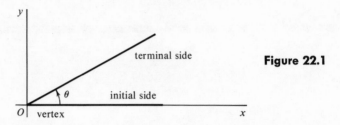

Figure 22.1

Figure 22.2(a) gives an example, where $\theta > 0$ in radians and Figure 22.2(b) has $\theta < 0$ in radians. Note that the terminal side of θ intersects C at the point $P(t)$ and then θ has measure t radians.

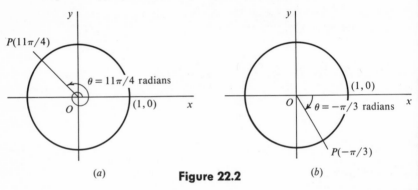

(a)

Figure 22.2

(b)

Another way of defining the radian measure of an angle is to let θ be a central angle of C, i.e., with vertex at the origin, the center of C. Then θ has radian measure one, if and only if θ subtends an arc of unit length on C.

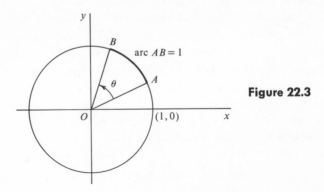

Figure 22.3

In either of these cases the end result is the same. A straight angle is thus an angle of π radians; the angle made by the x and y axes is $\pi/2$ radians; the angle between the bisector of the first quadrant and the positive x axis is $\pi/4$ radians; and so on.

Another type of angle measure is obtained by defining a straight angle to be an angle of 180 *degrees*, written 180°. Using the fact that there are π radians in a straight angle, it follows that there are $(180/\pi)°$ in an angle of one radian, and an angle of one degree (1°) is an angle of $\pi/180$ radians. Strictly speaking, it is not correct to write $1° = \pi/180$ radians, or $a° = b$ radians, because the units are different. However, one usually equates an angle with its measure and then no confusion need arise in saying, for example, that $\pi/2$ radians $= 90°$, 2π radians $= 360°$, $\pi/3$ radians $= 60°$, etc.

We now turn to the definition of $T(\alpha)$ for α an angle and T a trigonometric function.

Definition 22.1. Let α be an angle, and let t be the measure of α in radians. Then if T is any trigonometric function, define

$$T(\alpha) = T(t).$$

As every angle α has associated with it a radian measure, $T(\alpha)$ is defined for every angle α. Using Definition 22.1 and other definitions we may now write, for example,

$$\sin 45° = \sin \frac{\pi}{4} = \frac{1}{\sqrt{2}} = \frac{\sqrt{2}}{2}, \cos 60° = \cos \frac{\pi}{3} = \frac{1}{2},$$

and
$$\tan 240° = \tan \frac{4\pi}{3} = \sqrt{3}.$$

Other examples are found in the exercises below. We would like to re-emphasize that $T(a°)$ or $T(b$ radians$)$ means $T(\alpha)$, where α has measure $a°$ or b radians.

From the facts that the trigonometric functions are periodic and that an angle of π radians is equal to an angle of 180°, it follows that $T(\alpha) = T(\alpha + 2n\pi$ radians$) = T(\alpha + n \cdot 360°)$ for T the sine, cosine, secant, or cosecant, α any angle, and n any integer. For the tangent and cotangent it follows that

$$\tan \alpha = \tan (\alpha + n\pi \text{ radians}) = \tan (\alpha + n \cdot 180°)$$

and
$$\cot \alpha = \cot (\alpha + n\pi \text{ radians}) = \cot (\alpha + n \cdot 180°),$$

where α is an arbitrary angle and n any integer.

Recall that, in section **10**, the concept of reference number was introduced (Definition 10.2). This made it possible to find the values of the trigonometric functions at real numbers in terms of the same functions evaluated at numbers between 0 and $\pi/2$, e.g., $\tan 7\pi/6 = \tan \pi/6$. In a manner similar to that of Definition 10.2, reference angles are defined as follows:

Definition 22.2. Let α be an angle, $0 < \alpha < 360°$, $\alpha \neq 90°$, $\alpha \neq 180°$, $\alpha \neq 270°$. Then one of the angles, $180° - \alpha$, $\alpha - 180°$, or $360° - \alpha$, is between $0°$ and $90°$ and is called the *reference angle* associated with α. If α is an angle, θ its reference angle, and T a trigonometric function, then $|T(\alpha)| = T(\theta)$.

Example 22.3. (a) If $\alpha = 133°$, then its reference angle is $180° - 133° = 47°$. Thus, $\sin 133° = \sin 47°$, and from Table I, $\sin 47° = 0.7314$, so $\sin 133° = 0.7314$.

(b) Using the periodicity of the functions and reference angles we find such values as $\sin 2350°$. For division of $2350°$ by $360°$ gives $2350° = 6 \cdot 360° + 190°$. Thus, $\sin 2350° = \sin (6 \cdot 360° + 190°) = \sin 190°$. As the reference angle of $190°$ is $10°$, and $190°$ is an angle with its terminal side in the third quadrant, $\sin 2350° = -\sin 10°$. From Table I (page 153) $\sin 10° = 0.1736$. Therefore $\sin 2350° = -0.1736$.

Example 22.4. As an angle of $1°$ is the same as an angle of $\pi/180$ radians, it follows that an angle of $30°$ is an angle of $\pi/6$ radians, for example. As an angle of one radian is an angle of $(180/\pi)°$, it follows that an angle of $\pi/4$ radians is an angle of

$$\frac{\pi}{4}\left(\frac{180}{\pi}\right)^° = 45°.$$

One radian is approximately $57.30°$, to the nearest hundredth of a degree.

As a final result in this section, it should be pointed out that, from Definition 22.1 and the definitions of the trigonometric functions, all of the identities are now valid, if the number is replaced by an angle. We list them below as a review, where α and β are arbitrary angles.

1. $\sin^2 \alpha + \cos^2 \alpha = 1.$

2. $1 + \tan^2 \alpha = \sec^2 \alpha.$

3. $1 + \cot^2 \alpha = \csc^2 \alpha.$

4. $\csc \alpha = \dfrac{1}{\sin \alpha}.$

5. $\sec \alpha = \dfrac{1}{\cos \alpha}.$

6. $\tan \alpha = \dfrac{\sin \alpha}{\cos \alpha}.$

7. $\cot \alpha = \dfrac{\cos \alpha}{\sin \alpha}.$

8. $\cot \alpha = \dfrac{1}{\tan \alpha}.$

9. $\sin (\alpha + \beta) = \sin \alpha \cos \beta + \cos \alpha \sin \beta.$

10. $\sin (\alpha - \beta) = \sin \alpha \cos \beta - \cos \alpha \sin \beta.$

11. $\cos (\alpha + \beta) = \cos \alpha \cos \beta - \sin \alpha \sin \beta.$

12. $\cos (\alpha - \beta) = \cos \alpha \cos \beta + \sin \alpha \sin \beta.$

13. $\tan (\alpha + \beta) = \dfrac{\tan \alpha + \tan \beta}{1 - \tan \alpha \tan \beta}.$

14. $\tan (\alpha - \beta) = \dfrac{\tan \alpha - \tan \beta}{1 + \tan \alpha \tan \beta}.$

15. $\sin (-\alpha) = -\sin \alpha.$

16. $\cos (-\alpha) = \cos \alpha.$

17. $\sin \left(\dfrac{\pi}{2} - \alpha\right) = \sin (90° - \alpha) = \cos \alpha.$

18. $\cos \left(\dfrac{\pi}{2} - \alpha\right) = \cos (90° - \alpha) = \sin \alpha.$

19. $\cos 2\alpha = \cos^2 \alpha - \sin^2 \alpha$

$\qquad\quad = 2 \cos^2 \alpha - 1$

$\qquad\quad = 1 - 2 \sin^2 \alpha.$

20. $\sin 2\alpha = 2 \sin \alpha \cos \alpha.$

21. $\tan 2\alpha = \dfrac{2 \tan \alpha}{1 - \tan^2 \alpha}.$

22. $\cos \dfrac{\alpha}{2} = (\text{sign}) \sqrt{\dfrac{1 + \cos \alpha}{2}}.$

23. $\sin \dfrac{\alpha}{2} = (\text{sign}) \sqrt{\dfrac{1 - \cos \alpha}{2}}.$

24. $\tan \dfrac{\alpha}{2} = (\text{sign}) \sqrt{\dfrac{1 - \cos \alpha}{1 + \cos \alpha}}$

$\qquad = \dfrac{1 - \cos \alpha}{\sin \alpha}$

$\qquad = \dfrac{\sin \alpha}{1 + \cos \alpha}.$

Note. Each degree is divided into 60 equal parts called minutes, and each minute is divided into 60 equal parts called seconds. For our purposes of computation we will only be concerned with accuracy of angular measure to the nearest degree.

EXERCISES

1. Change the following from degrees to radians. Leave answers in terms of π.
   (a) 60°   (c) 90°   (e) 135°   (g) 118°   (i) 15°
   (b) 75°   (d) 120°   (f) 150°   (h) 36°   (j) 27°.

2. Change the following from radians to degrees. Leave answers in fraction form.

   (a) $\dfrac{\pi}{3}$ (c) $\dfrac{\pi}{12}$ (e) $\dfrac{\pi}{7}$ (g) $\dfrac{\pi}{10}$ (i) $\dfrac{5\pi}{14}$

   (b) $\dfrac{\pi}{9}$ (d) $\dfrac{\pi}{5}$ (f) 3   (h) $\dfrac{3\pi}{11}$ (j) $\dfrac{12\pi}{11}$.

3. Complete the following table:

Angle θ	sin θ	cos θ	tan θ	cot θ	sec θ	csc θ
0°						
30°						
45°						
60°						
90°						
120°						
135°						
150°						
180°						

4. Evaluate the following.

(a) tan 930° (c) sec (−2730°), (e) cos 750°

(b) cot 3150° (d) sin (−9405°) (f) sin (−1500°).

5. Use periodicity, reference angles, and Table I (page 153) to evaluate the following.

(a) sin 237° (c) tan (−430°) (e) cos (−1600°)

(b) cos 1234° (d) sin (−3940°) (f) tan (−5935°).

6. Find the exact value of each of the following, using the identities above and the table of Problem 3.

(a) sin 75° (c) sin $22\frac{1}{2}°$ (e) cos $67\frac{1}{2}°$

(b) cos 105° (d) tan 165° (f) cos $7\frac{1}{2}°$.

7. Find cos 75° by two different methods. Show that the values are the same.

8. If $A + B + C = 180°$, show:

(a) $\sin (B + C) = \sin A$

(b) $\sin \frac{1}{2}(B + C) = \cos \frac{1}{2}A$.

9. If tan 35° = a, find:

(a) $\dfrac{\tan 145° - \tan 125°}{1 + \tan 145° \tan 125°}$

(b) $\dfrac{\tan 215° - \tan 125°}{\tan 235° + \tan 325°}$.

23. Polar Coordinates

Let $P(x,y)$ be any point in the plane and θ an angle in standard position with terminal side passing through P. Point P' on the unit circle through which the terminal side of θ passes has coordinates (cos θ, sin θ). Let $r = |OP|$, i.e., $r = \sqrt{x^2 + y^2}$. The points P and P' are both on the terminal side of θ and perpendiculars are dropped from each point to the x axis, the points of intersection on the x axis being labeled Q and Q', respectively. The resulting triangles using the reference angle of θ, triangles OPQ and $OP'Q'$, are similar. Hence, the corresponding parts are proportional, and it follows that

$$\frac{|OQ'|}{|OP'|} = \frac{|OQ|}{|OP|} \quad \text{and} \quad \frac{|P'Q'|}{|OP'|} = \frac{|PQ|}{|OP|},$$

or

$$|\cos \theta| = \frac{|x|}{r} \quad \text{and} \quad |\sin \theta| = \frac{|y|}{r},$$

as $|OP'| = 1$ and $|OP| = r$. As x and cos θ have the same sign, we may

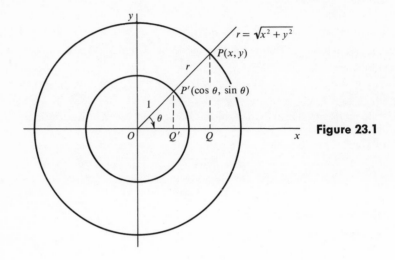

Figure 23.1

now write $x = r \cos \theta$. Similarly, it follows that $y = r \sin \theta$. Using this discussion, we have the following definition.

Definition 23.1. The polar coordinates of a point $P(x,y)$ are defined by the equations

$$x = r \cos \theta \quad \text{and} \quad y = r \sin \theta,$$

where $\sqrt{x^2 + y^2} = r$ and θ is an angle in standard position with terminal side passing through $P(x,y)$. The polar coordinates will be given by $[r,\theta]$.

It should be observed that the polar coordinates of a point are not unique. For

$$[r,\theta] = [r,\theta + k \cdot 360°] = [r,\theta + m \cdot 2\pi \text{ radians}],$$

where k and m are arbitrary integers.

This section is the basis for an alternate way of defining trigonometric functions of angles. This is done by starting with an angle θ in standard position and selecting an arbitrary point $P(x,y)$ on the terminal side of θ. Then, if $r = \sqrt{x^2 + y^2}$, define $\cos \theta = x/r$, $\sin \theta = y/r$, $\tan \theta = y/x$, $x \neq 0$, $\cot \theta = x/y$, $y \neq 0$, $\csc \theta = r/y$, $y \neq 0$ and $\sec \theta = r/x$, $x \neq 0$.

It is easy to make a change in coordinates from polar to rectangular coordinates by merely using the equations of Definition 23.1. For example, given $P[2,25°]$ we have $x = 2 \cos 25°$ and $y = 2 \sin 25°$. Then using

Table I, sin $25° \doteq 0.4226$ and cos $25° \doteq 0.9063$, and so $x = 1.8126$ and $y = 0.8432$ or $P(x,y) = (1.8126,0.8432)$. To go the other way, change from rectangular to polar coordinates, requires more work. For example, if $P(x,y) = (1,2)$, then $r = \sqrt{1^2 + 2^2} = \sqrt{1 + 4} = \sqrt{5}$. From Definition 23.1 it follows that cos $\theta = 1/\sqrt{5}$ and sin $\theta = 2/\sqrt{5}$. Using 0.4472 as an approximation for $1/\sqrt{5}$, cos $\theta \doteq 0.4472$ and sin $\theta \doteq 0.8944$, and from Table I, $\theta \doteq 63\frac{1}{2}°$. Thus, $[r,\theta] = [\sqrt{5},63\frac{1}{2}°]$.

EXERCISES

1. Write in rectangular coordinates.

(a) $[2,60°]$

(b) $\left[3,-\dfrac{\pi}{4} \right]$

(c) $[4,\pi]$

(d) $[\frac{1}{2},240°]$

(e) $[\frac{2}{3},-300°]$

(f) $[1750°]$

(g) $[2,-2925°]$

(h) $\left[\sqrt{2},27\dfrac{\pi}{4} \right]$.

2. Use Table I (page 153) to write in rectangular coordinates. Give answer to the nearest hundredth.

(a) $[4,58°]$

(b) $[3,26°]$

(c) $[4,-221°]$

(d) $[\sqrt{5},-100°]$.

3. Write in polar coordinates, using Table I (page 153) if necessary. Find to the nearest degree.

(a) $(1,1)$

(b) $(3,-1)$

(c) $(2,4)$

(d) $(-4,2)$

(e) $(-3,-3)$

(f) $(-2,-4)$

(g) $(1,5)$

(h) $(2,3)$

(i) $(2,-2)$.

24. Right Triangle Trigonometry, Solutions of Arbitrary Triangles

Let A, B, C be the vertices of a right triangle, with α an acute angle in standard position and such that the hypotenuse of the triangle is the terminal side of α. The vertex is thus at the origin. (See Figure 24.1 below.) Let $c = |AC|$ be the length of the hypotenuse, and $|AB| = a$,

Figure 24.1

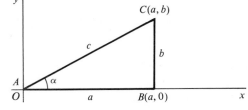

$|BC| = b$ the lengths of the other two sides. Then the coordinates of C are (a,b). By the Pythagorean theorem $a^2 + b^2 = c^2$, and by the definition of polar coordinates, $a = c \cos \alpha$ and $b = c \sin \alpha$. Hence, $\cos \alpha = a/c$ and $\sin \alpha = b/c$. As b is the length of the side opposite α and a the length of the side adjacent to α, one has in words,

$$\sin \alpha = \frac{|\text{side opposite } \alpha|}{|\text{hypotenuse}|}$$

and
$$\cos \alpha = \frac{|\text{side adjacent to } \alpha|}{|\text{hypotenuse}|}.$$

It is clear that the angles α and $\beta = (\pi/2) - \alpha$ can be interchanged. Thus, for any acute angle of a right triangle, the above relations are valid and serve as a means for computing the functions at these angles, knowing the lengths of the sides.

In any triangle, whether or not it is a right triangle, there are six parts: the three sides and the three angles. We say that we solve a triangle if, given some of these six parts, we are able to find the remaining parts. In general, if three parts are known, at least one of which is the length of a side, then the other three parts can be found when they exist. The combinations of permissible known parts are included in the following cases: (i) Two angles and a side are known. (ii) Two sides and the angle

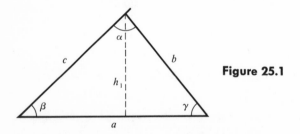

Figure 25.1

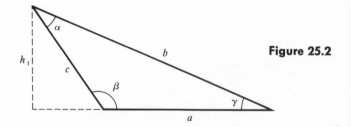

Figure 25.2

opposite one of them are known. (iii) Two sides and the included angle are known. (iv) The three sides are known. Cases (i) and (ii) are obtained by using the Law of Sines while cases (iii) and (iv) use the Law of Cosines. The Law of Sines will be derived first.

25. The Law of Sines

Let α, β, and γ be the three angles of a triangle with sides of length a, b, c opposite the respective angles. As shown in Figures 25.1 and 25.2, drop a perpendicular from the vertex of α to the side opposite it. Let h_1 represent the length of this perpendicular segment. As the figures indicate, there are two possible locations for this perpendicular segment— it is either interior or exterior to the triangle. In Figure 25.1, we have

$$\sin \beta = \frac{h_1}{c} \quad \text{and} \quad \sin \gamma = \frac{h_1}{b}.$$

In Figure 25.2,

$$\sin \beta = \sin (\pi - \beta) = \frac{h_1}{c} \quad \text{and} \quad \sin \gamma = \frac{h_1}{b}.$$

In either case it follows that $h_1 = c \sin \beta = b \sin \gamma$. Hence,

$$(25.1) \qquad \frac{\sin \beta}{b} = \frac{\sin \gamma}{c}.$$

By dropping a perpendicular from either β or γ to the side b or c, respectively, we conclude either

$$(25.2) \qquad \frac{\sin \alpha}{a} = \frac{\sin \gamma}{c},$$

$$(25.3) \quad \text{or} \qquad \frac{\sin \alpha}{a} = \frac{\sin \beta}{b}.$$

However, when either (25.2) or (25.3) is used with (25.1) we have

$$(25.4) \qquad \frac{\sin \alpha}{a} = \frac{\sin \beta}{b} = \frac{\sin \gamma}{c}.$$

This equation is known as the Law of Sines. We now illustrate its use.

Example 25.1. Solve the triangle in which $\alpha = 50°$, $\beta = 55°$, and $c = 25$.

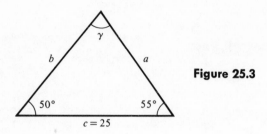

Figure 25.3

Solution. This is an example of case (i) of section **24**. First of all note that since $\alpha + \beta + \gamma = 180°$, $\gamma = 75°$. Now applying the Law of Sines gives

$$\frac{\sin \gamma}{c} = \frac{\sin \alpha}{a},$$

or $\qquad a = c\,\frac{\sin \alpha}{\sin \gamma} = 25\,\frac{\sin 50°}{\sin 75°} = 25\,\frac{0.7660}{0.9659} = 19.8.$

We also have

$$b = c\,\frac{\sin \beta}{\sin \gamma} = 25\,\frac{\sin 55°}{\sin 75°} = 25\,\frac{0.8192}{0.9659} = 21.2.$$

It should be pointed out that logarithms can be used quite easily with the Law of Sines. For example,

$$\log a = \log 25\,\frac{\sin 50°}{\sin 75°} = \log 25 + \log \sin 50° - \log \sin 75°,$$

and using the tables in the back of the book the student may compute first $\log a$ and then a.

Note. We wish to emphasize that the values given in the tables for $\sin \alpha$, $\log \sin \alpha$, or $\log a$ are only approximations. Hence, the use of the equal sign in the above relations is, strictly speaking, incorrect. However, for our purposes it will be sufficient to know that most of the entries are approximations, and we will continue to write equality.

Example 25.2. Solve the triangle with α, a, and b given.

Solution. This is the general form of case (ii) of section 24. The figure below is used as a guide.

Figure 25.4

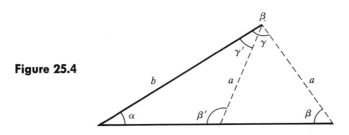

This is called the *ambiguous case*, for its solution leads to three possibilities: one triangle, two triangles, or no triangle. Geometrically, it should be clear from the diagram, that these possibilities arise. For using the point B as center and a as the radius of a circle, the number of times this circle intersects the horizontal leg of the angle α is the number of solutions.

Analytically, we have the following:

$$\frac{\sin \beta}{b} = \frac{\sin \alpha}{a},$$

or

$$\sin \beta = \frac{b}{a} \sin \alpha.$$

If $b/a \sin \alpha > 1$, there is no solution, for $\sin \beta \le 1$. If $b/a \sin \alpha = e \le 1$, then there is at least one solution. In particular, when $e = 1$, $\beta = 90°$ and there is one solution. In that case b is the length of the hypotenuse, $c = \sqrt{b^2 - a^2}$, and $\gamma = 90° - \alpha$. If $e < 1$, then there are two angles, $\beta < 90°$ and $\beta' = 180° - \beta$, between $0°$ and $180°$, such that $\sin \beta = \sin \beta' = e$. If both $\alpha + \beta < 180°$ and $\alpha + \beta' < 180°$, then there are two triangles. One has α, β, $180° - (\alpha + \beta)$ and sides a, b, and c. The second has angles α, β', $180° - (\alpha + \beta')$ and sides a, b, and c'. As in Example 25.1, logarithms may be used to find the length(s) of the side(s) c (and c').

It should be observed that the Law of Sines can be used to solve some right triangles, although the relations given in section **24** are usually used.

EXERCISES

Solve the following right triangles. c is the length of the hypotenuse.

1. $a = 5$, $b = 12$.

2. $b = 7$, $c = 25$.

3. $a = 13$, $b = 9$.

4. $\alpha = 40°$, $b = 20$.

5. $\beta = 65°$, $a = 10$.

6. $\alpha = 35°$, $c = 100$.

Solve Problems 7–16 using the Law of Sines.

7. $\alpha = 40°$, $\beta = 35°$, $c = 100$.

8. $\alpha = 50°$, $\beta = 50°$, $c = 100$.

9. $\alpha = 30°$, $\beta = 80°$, $c = 100$.

10. $\alpha = 35°$, $\beta = 65°$, $c = 100$.

11. $\alpha = 45°$, $\beta = 60°$, $c = 100$.

12. $\alpha = 45°$, $a = 100$, $b = 75$.

13. $\alpha = 45°$, $a = 25$, $b = 75$.

14. $\alpha = 30°$, $a = 100$, $b = 50$.

15. $\alpha = 60°$, $a = 100$, $b = 50$.

16. $\alpha = 50°$, $a = 50$, $b = 50$.

17. A pathway cuts across a rectangular lot 30 feet from the corner and comes out 45 feet from the same corner. How long is the pathway?

18. A 20-foot ladder is placed against a vertical wall such that its foot is 6 feet from the wall. What angle does the ladder make with the wall and how high on the wall does the ladder reach?

19. A surveyor wishes to measure the width of a river. He stands at a point A with point B opposite him on the other side of the river. From A he walks 300 feet downstream to a point C. The line CA makes an angle of 35° with CB. How wide is the river?

20. A man stands 200 feet from a building and the angle of elevation of the top of the building, i.e., the angle made by his line of sight with the horizontal, is 35°. When he moves 100 feet farther from the building, the angle of elevation is 25°. How tall is the building?

21. A cannon having an elevation of 40° has a range of 10,000 yards. Assuming the trajectory of a shell to be very nearly an isosceles triangle, what is the altitude of the shell at its highest point and what is the total distance it travels?

22. Two men stand 200 feet apart with a flagpole situated directly between them. If the angles of elevation of the top of the flagpole are 30° and 25°, how tall is the pole and how far is each man from it?

23. A tower 100-feet-tall is located at the top of a hill. At a point 500 feet down the hill the angle between the surface of the hill and the line of sight to the top of the tower is 10°. Find the inclination of the hill to a horizontal plane.

24. A point of land is located 20 miles northeast of a dock. A ship leaves the dock at 10 A.M. traveling east at 12 mph. At what time is the ship 15 miles from the point?

25. Trees A and B are 500 feet apart on the same side of a swamp opposite a tree C. If the angle between lines AB and AC is 70° and that between AB and BC is 35°, find the distances from A to C and from B to C.

26. The Law of Cosines

To derive the Law of Cosines, let α, β, and γ be the three angles of a triangle, and a, b, and c the lengths of the sides opposite the respective angles. As in the Law of Sines there are two possible cases for the triangles: one in which all the angles are $\leq$ 90°, and another with one angle $>$ 90°, and they are handled separately. (See Figures 26.1 and 26.2.)

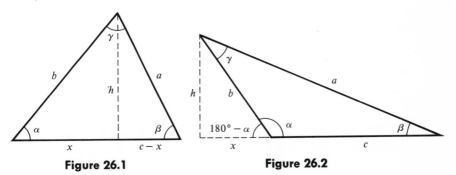

Figure 26.1   Figure 26.2

If all the angles are $\leq$ 90°, drop a perpendicular from the angle γ to its opposite side, dividing that side into segments of length x and $c - x$. Then, if h is the altitude of the triangle,

$$b^2 = h^2 + x^2,$$

and $$a^2 = h^2 + (c - x)^2.$$

Then, as $h^2 = b^2 - x^2$,

$$a^2 = b^2 - x^2 + (c - x)^2,$$
$$= b^2 - x^2 + c^2 - 2cx + x^2,$$
$$= b^2 + c^2 - 2cx.$$

Now, $\cos \alpha = x/b$, or $x = b \cos \alpha$. Hence,

(26.1) $$a^2 = b^2 + c^2 - 2bc \cos \alpha.$$

In a similar manner, it follows that

(26.2) $$b^2 = a^2 + c^2 - 2ac \cos \beta,$$

(26.3) and $$c^2 = a^2 + b^2 - 2ab \cos \gamma.$$

These equations can also be written in the forms

(26.4) $$\cos \alpha = \frac{b^2 + c^2 - a^2}{2bc},$$

(26.5) $$\cos \beta = \frac{a^2 + c^2 - b^2}{2ac},$$

(26.6) $$\cos \gamma = \frac{a^2 + b^2 - c^2}{2ab}.$$

Any one of these equations could be called the Law of Cosines, as each of them can be obtained from the others by substitution and either multiplication or division.

In the case $\alpha > 90°$, we have the Figure 26.2. As in the previous case, drop a perpendicular from the angle γ to the extended opposite side and let x be the length of the segment cut by the perpendicular. Then if h is the altitude of the triangle,

$$b^2 = h^2 + x^2,$$

and $$a^2 = h^2 + (c + x)^2,$$

and as $h^2 = b^2 - x^2$,

$$\begin{aligned} a^2 &= b^2 - x^2 + (c + x)^2 \\ &= b^2 - x^2 + c^2 + 2cx + x^2 \\ &= b^2 + c^2 + 2cx. \end{aligned}$$

Now $\cos (180° - \alpha) = x/b$ and $\cos \alpha = -x/b$ or $x = -b \cos \alpha$. Thus,

$$a^2 = b^2 + c^2 - 2bc \cos \alpha,$$

which is Equation (26.1). The remaining equations are also the same, so it follows that the Law of Cosines is valid for all triangles.

Consider now the following examples.

Example 26.1. Solve the triangle in which $a = 20$, $b = 25$, and $\gamma = 60°$.

Figure 26.3

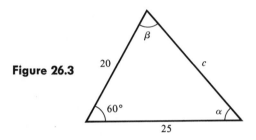

Solution. This is an example of case (iii) of section **24**. Applying Equation (26.3) yields

$$c^2 = a^2 + b^2 - 2ab \cos \gamma$$
$$= 20^2 + 25^2 - 2 \cdot 25 \cdot 20 \cdot \tfrac{1}{2}$$
$$= 400 + 625 - 500 = 525.$$

Hence, $c = \sqrt{525} = 5\sqrt{21}$.

To find the angles α and β, apply the Law of Sines to get

$$\frac{\sin \alpha}{a} = \frac{\sin \gamma}{c}.$$

Then

$$\sin \alpha = \frac{20}{5\sqrt{21}} \sin 60° = \frac{4}{\sqrt{21}} \frac{\sqrt{3}}{2} = \frac{2\sqrt{3}}{\sqrt{21}} = 2\sqrt{\frac{3}{21}},$$

or $\sin \alpha = 2\sqrt{7}/7$. Thus, $\sin \alpha = 0.756$, approximately, and $\alpha = 49°$ to the nearest degree. $\beta = 180° - (\alpha + \gamma) = 180° - 109° = 71°$.

Example 26.2. Solve the triangle in which $a = 20$, $b = 25$, and $c = 30$.

Solution. Applying Equation (26.4) gives

$$\cos \alpha = \frac{b^2 + c^2 - a^2}{2bc}$$

$$= \frac{25^2 + 30^2 - 20^2}{2 \cdot 25 \cdot 30} = \frac{625 + 900 - 400}{1500}$$

$$= \frac{1125}{1500} = 0.750.$$

Thus, to the nearest degree, $\alpha = 41°$.

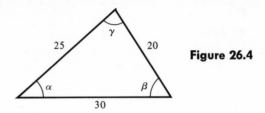

Figure 26.4

To find either β or γ, the Law of Cosines is used again rather than the Law of Sines, for the use of the Law of Sines would compound the error that arises in obtaining α. Thus,

$$\cos \beta = \frac{a^2 + c^2 - b^2}{2ac}$$

$$= \frac{20^2 + 30^2 - 25^2}{2 \cdot 20 \cdot 30} = \frac{400 + 900 - 625}{1200}$$

$$= \frac{725}{1200} = 0.604 \text{ to 3 decimal places.}$$

Then, $\beta = 53°$, to the nearest degree, and having obtained α and β, $\gamma = 180° - (\alpha + \beta) = 180° - 94° = 86°$.

EXERCISES

Solve the following triangles:

1. a = 50, b = 100, c = 125.

2. a = 100, b = 100, c = 125.

3. a = 100, b = 200, c = 225.

4. $a = 15, \quad b = 9, \quad \gamma = 120°.$

5. $b = 20, \quad c = 15, \quad \alpha = 67°.$

6. $a = 42, \quad c = 37, \quad \beta = 79°.$

7. $a = 15, \quad c = 26, \quad \beta = 112°.$

8. Three circles of radii 110, 150, and 200 feet, respectively, are tangent to each other externally. Find the angles of the triangle formed by joining the centers of the circles.

9. A parallelogram has adjacent sides 10 inches and 15 inches long. If the shorter diagonal is 12 inches long, what is the length of the longer diagonal?

10. A man stands at a point A, 200 feet from point B and 150 feet from point C. If the angle between AB and AC is 40°, how far apart are B and C?

11. If two forces act at a point, the magnitude of their sum or resultant is the length of the diagonal of a parallelogram where the lengths of the adjacent sides are the magnitudes of the two respective forces. If forces of 20 and 35 lbs. act at an angle of 30° to each other, what is the resultant force and what is its angle to each of the two forces?

12. Forces of 30 and 40 lbs., acting at a point, have a resultant of 55 lbs. What is the angle between the two forces?

27. Area Formulas

The area of a triangle is given by the formula

$$K = \tfrac{1}{2}bh,$$

where h is the altitude and b the length of the base of the triangle. It is the purpose of this section to derive other formulas which may be used when the altitude is not known. The Law of Sines and the Law of Cosines will both be used in the developments. The figure below will serve as a guide.

Figure 27.1

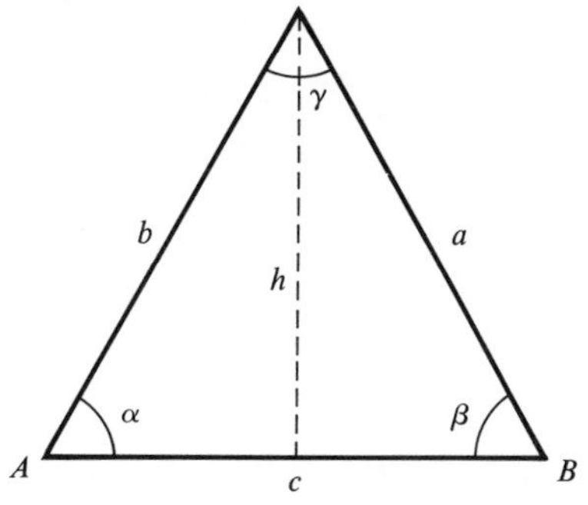

The area of the triangle ABC is given by

$$K = \tfrac{1}{2}ch.$$

Now, $h = b \sin \alpha$, and so it follows that $K = \tfrac{1}{2}bc \sin \alpha$. But $h = a \sin \beta$ also, and so $K = \tfrac{1}{2}ac \sin \beta$. One can also obtain $K = \tfrac{1}{2}ab \sin \gamma$. Thus any one of the equations below is a formula for the area of the triangle ABC.

(27.1) $$K = \begin{cases} \tfrac{1}{2}ab \sin \gamma \\ \tfrac{1}{2}ac \sin \beta \\ \tfrac{1}{2}bc \sin \alpha. \end{cases}$$

From the Law of Sines it follows that

$$b = \frac{a \sin \beta}{\sin \alpha}, \quad a = \frac{c \sin \alpha}{\sin \gamma}, \quad \text{and} \quad c = \frac{b \sin \gamma}{\sin \beta}.$$

Hence, the equations of (27.1) become

(27.2) $$K = \begin{cases} \tfrac{1}{2}a^2 \dfrac{\sin \beta \sin \gamma}{\sin \alpha} \\[2mm] \tfrac{1}{2}c^2 \dfrac{\sin \alpha \sin \beta}{\sin \gamma} \\[2mm] \tfrac{1}{2}b^2 \dfrac{\sin \alpha \sin \gamma}{\sin \beta}. \end{cases}$$

In each of the formulas there are only three parts of the triangle necessary to find the area. One more formula can also be obtained, for the area in terms of the three sides.

Using $K = \tfrac{1}{2}ab \sin \gamma$ from Equation (27.1) and squaring both sides gives

$$\begin{aligned} K^2 &= \tfrac{1}{4}a^2b^2 \sin^2 \gamma \\ &= \tfrac{1}{4}a^2b^2 (1 - \cos^2 \gamma) \\ &= \tfrac{1}{4}a^2b^2 \left[1 - \left(\frac{a^2 + b^2 - c^2}{2ab} \right)^2 \right], \text{ by the Law of Cosines.} \\ &= \tfrac{1}{4}a^2b^2 \left[\frac{4a^2b^2 - (a^2 + b^2 - c^2)^2}{4a^2b^2} \right] \\ &= \tfrac{1}{16}[4a^2b^2 - (a^2 + b^2 - c^2)^2]. \end{aligned}$$

Factoring this last expression as the difference of two squares gives

$$K^2 = \tfrac{1}{16}[2ab - (a^2 + b^2 - c^2)][2ab + (a^2 + b^2 - c^2)]$$

$$= \tfrac{1}{16}[c^2 - (a^2 - 2ab + b^2)][(a^2 + 2ab + b^2) - c^2]$$

$$= \tfrac{1}{16}[c^2 - (a - b)^2][(a + b)^2 - c^2].$$

Factoring each of these expressions as the difference of two squares then yields

$$K^2 = \tfrac{1}{16}[c - (a - b)][c + (a - b)][(a + b) - c][(a + b) + c]$$

$$= \left(\frac{b + c - a}{2}\right)\left(\frac{a + c - b}{2}\right)\left(\frac{a + b - c}{2}\right)\left(\frac{a + b + c}{2}\right).$$

Introducing the number

$$s = \tfrac{1}{2}(a + b + c),$$

the semiperimeter of the triangle allows us to write

$$\frac{b + c - a}{2} = s - a,$$

$$\frac{a + c - b}{2} = s - b,$$

$$\frac{a + b - c}{2} = s - c.$$

Thus, $K^2 = s(s - a)(s - b)(s - c)$, and extracting square roots from both sides gives us Heron's formula

(27.3) $$K = \sqrt{s(s - a)(s - b)(s - c)}.$$

EXERCISES

1. Find the area in each of the problems of the Exercises of sections **25** and **26**.

2. Using Heron's formula, derive the formula $K = (\sqrt{3}/4)a^2$ for area of an equilateral triangle whose sides have length a.

3. If a triangle has sides of lengths a, b, and c, show that $K = rs$, where r is the radius of the circle inscribed in the triangle and $s = \frac{1}{2}(a + b + c)$. Note that then $r = \dfrac{\sqrt{(s - a)(s - b)(s - c)}}{s}$. Figure 27.2 may be helpful.

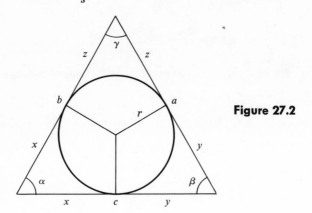

Figure 27.2

28. Arc Length, Angular and Linear Velocity

Let $C_r = \{(x,y) : x^2 + y^2 = r^2\}$, i.e., C_r is the circle of radius r and with its center at the origin. Let θ be a central angle whose positive measure is θ radians, and let s be the length of arc on C_r subtended by θ. (See Figure 28.1 below.) From the discussion of radian measure in section **22**

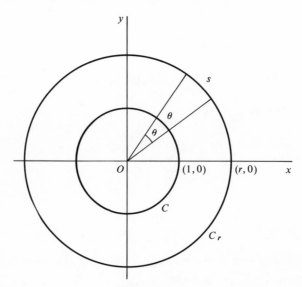

Figure 28.1

it is clear that on C the length of arc subtended by this central angle is θ. Geometrically the proportion

$$\frac{s}{\theta} = \frac{r}{1}$$

must hold, i.e., the ratio of the length of the arc on C_r to the length of the arc on C is the same as the ratio of the radius r of C_r to the radius 1 of C. Hence,

(28.1) $s = r\theta$

is a formula for the length of the arc subtended by the central angle of θ radians in a circle of radius r.

Now let a particle be placed at a point P on C_r, and suppose the particle moves on C_r for a time interval t units in length from the point P to the point Q on C_r. Suppose further that θ is the total central angle sub-

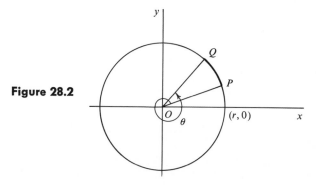

Figure 28.2

tending the arc through which the particle moves. Then, from above, the total distance traveled by the particle is $s = r\theta$. Dividing both sides by t gives the formula

$$\frac{s}{t} = r\frac{\theta}{t}.$$

Setting $v = \dfrac{s}{t}$ and $\omega = \dfrac{\theta}{t}$, gives

(28.2) $v = r\omega.$

v is to be interpreted as "linear velocity," distance per unit of time, and ω as "angular velocity," the number of radians per unit of time generated by the segment through O and the particle in going from P to Q.

Example 28.1. A wheel of radius 8 inches turns at the rate of 25 revolutions per second: (a) Find the linear and angular velocity of a point on the rim of the wheel. (b) Find the distance traveled by a point on the wheel in 2 seconds.

Solution. As the formulas all involve angles in radian measure, it is necessary to write 25 revolutions as $25 \cdot 2\pi = 50\pi$ radians. Hence, ω, the angular velocity, is 50π radians/second. As $v = r\omega$ and $r = \frac{2}{3}$ feet, $v = \frac{2}{3}(50\pi) = (100\pi/3)$ feet/second. Using Equation (28.1) for (b) with $\theta = 100\pi$ radians gives $s = \frac{2}{3}(100\pi)$ feet $= (200\pi/3)$ feet.

EXERCISES

1. A circle has radius 25 inches. What is the length of arc subtended by a central angle of $\frac{1}{3}$ radian?

2. What is the length of arc subtended by a central angle of 40° in a circle of radius 20 inches?

3. What is the central angle subtended by an arc of length $1\frac{1}{2}$ feet on a circle of radius 30 inches?

4. The minute hand of a clock is 12 inches long. How far does the tip move during 20 minutes? 25 minutes?

5. A car is moving at the rate of 30 mph along a circular roadway of radius 1000 feet. Through what angle does it turn in one minute?

6. The end of a 36-inch pendulum describes an arc of 4 inches. Through what angle does it swing?

7. An automobile tire is 30 inches in diameter. How fast, in rpm, does the wheel turn when the auto maintains a speed of 30 mph?

8. If an automobile wheel 30 inches in diameter rotates at 600 rpm, what is the speed of the car in mph?

9. What is the diameter of a pulley which is driven at 360 rpm by a belt moving at 40 feet/second?

10. A tricycle has back wheels 8 inches in diameter, the front wheel 18 inches in diameter, and the pedals are 6 inches long. If a boy peddles at the rate of 30 rpm, how far will the tricycle travel in 5 minutes? What is the angular

velocity of the back wheels in radians/second? How fast is the tricycle going in feet/second? What is the length of arc swept out by the pedal in 5 minutes?

11. A tractor has a belt pulley 10 inches in diameter, and the pulley on a blower is 12 inches in diameter. The pulley on the tractor runs at a speed of 800 rpm and drives a belt to the blower. What is the angular velocity in radians/minute of the pulley on the blower, and what is the linear velocity of the belt?

29. Area of Sectors and Segments of Circles

Associated with every pair of radii that do not constitute a diameter of a circle are the two arcs joining their end points. The smaller arc will be called the minor arc and the larger one the major arc.

A sector of a circle is a part of a circle determined by a central angle θ, the radii forming θ, and the arc subtended by it. (See Figure 29.1.) Here, it is assumed that θ is the positive radian measure of the angle also.

Figure 29.1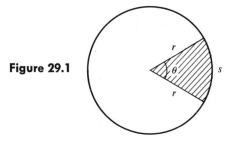

Let K be the area of a sector having θ as its central angle in a circle of radius r. Geometrically, K is proportional to θ,

$$K = k\theta,$$

where k is a constant of proportionality to be determined. Then using the fact that when $\theta = 2\pi$, the area of the circle is πr^2, it follows that $k = \frac{1}{2}r^2$. Thus the formula for the area of a sector is

(29.1) $K = \frac{1}{2}r^2\theta.$

A segment of a circle is a part of a circle bounded by a chord and the arc cut by the chord. (The minor arc will be used here. See Figure 29.2.) As the arc subtends a central angle, the area K of the segment is the area determined by θ minus the area of the triangle cut off by the chord. Thus using Equation (27.1) for the area of the triangle,

$$K = \tfrac{1}{2}r^2\theta - \tfrac{1}{2}r^2 \sin \theta,$$

(29.2)   or $$K = \tfrac{1}{2}r^2 (\theta - \sin \theta).$$

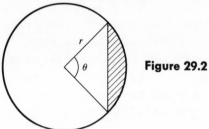

Figure 29.2

Example 29.1.   Find the area of the sector and segment subtended by a central angle of 20° in a circle of radius 9 inches.

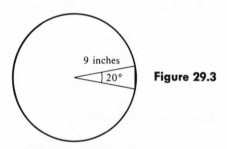

9 inches

20°   **Figure 29.3**

Solution.   As the formulas for the areas of sectors and segments require the measurement of the central angles in terms of radians, it is necessary to write 20° as

$$\frac{\pi}{180}\cdot 20 = \frac{\pi}{9} \text{ radians.}$$

Then the area of the sector is

$$K = \tfrac{1}{2}r^2\theta = \tfrac{1}{2}(9)^2\left(\frac{\pi}{9}\right) = \frac{9\pi}{2} \text{ sq. inches.}$$

The area of the segment is

$$K = \tfrac{1}{2}r^2\,(\theta - \sin\theta),$$

$$= \tfrac{1}{2}(9)^2\left(\frac{\pi}{9} - \sin 20°\right),$$

$$= \tfrac{1}{2}\cdot 81\left(\frac{\pi}{9} - 0.342\right),$$

$$= \tfrac{1}{2}(9\pi) - 81(0.171),$$

$$= 9\pi/2 - 13.85 \text{ sq. inches.}$$

EXERCISES

1. Find the areas of the sector and segment subtended by a central angle of $\pi/4$ radians in a circle of radius 6 inches.

2. Find the areas of the sector and segment subtended by a central angle of 50° in a circle of radius 3 inches.

3. Find the areas of the sector and segment subtended by a central angle of 105° in a circle of radius 5 inches.

4. If a sector having area $24\pi^2$ sq. inches is subtended by a central angle of $\pi/3$ radians, what is the radius of the circle?

5. If a sector has area 12π sq. inches in a circle of radius 4 inches, what is the central angle subtending the sector?

6. If a sector has area 25 sq. inches in a circle of radius 10 inches, what is the central angle subtending the sector?

5

INVERSE FUNCTIONS
AND TRIGONOMETRIC EQUATIONS

30. The Inverse of a Function

Recall that a (real) function f is a set of ordered pairs of (real) numbers (a,b) such that, if $(x,y) \in f$ and $(x,z) \in f$ then $y = z$. Associated with every function f is a set of ordered pairs which we will call the converse of f and define as follows.

Definition 30.1. Let f be a function. Then the *converse* of f, f^c, is defined by

$$f^c = \{(x,y) : (y,x) \in f\}.$$

Some examples will help to illustrate the definition.

Example 30.2. Let $f = \{(1,2), (2,4), (3,2)\}$. Then $f^c = \{(2,1), (4,2), (2,3)\}$.

Example 30.3. Let $g = \{(x,y) : y = x^2\}$. Then $g^c = \{(y,x) : y = x^2\}$ $= \{(x,y) : x = y^2\}$.

Example 30.4. Let $g = \{(x,y) : y = 2x\}$. Then $g^c = \{(y,x) : y = 2x\}$ $= \{(x,y) : x = 2y\}$.

We wish to emphasize that the choice of symbols used to represent the first or second elements of the ordered pairs is immaterial.

It should be observed that, in Examples 30.2 and 30.3, f^c and g^c are not functions, and, in Example 30.4, g^c is a function. Also, in Example 30.4, it is possible to compute $g \circ g^c$ and $g^c \circ g$. They are, using Definition 6.14,

$$g \circ g^c = \{(x,z): \text{ for some } y, \quad (x,y) \in g^c \text{ and } (y,z) \in g\}$$
$$= \{(x,z): \text{ for some } y, \quad x = 2y \text{ and } z = 2y\}$$
$$= \{(x,z) : x = z\}.$$

$$g^c \circ g = \{(x,z): \text{ for some } y, \quad (x,y) \in g \text{ and } (y,z) \in g^c\}$$
$$= \{(x,z): \text{ for some } y, \quad y = 2x \text{ and } y = 2z\}$$
$$= \{(x,z) : x = z\}.$$

Thus, it is seen that in this case $g \circ g^c = g^c \circ g = I$, the identity function. The case of Example 30.4 leads to the following definition.

Definition 30.5. Let f be a function and f^c the converse of f. f^c is called the *inverse* of f and indicated by f^{-1}, if and only if f^c is a function. Then $f \circ f^{-1} = f^{-1} \circ f = I$, the identity function.

As the above examples illustrate, not every function has an inverse even though every function has a converse. In fact, it is a very special type of function which does have an inverse. This is the type defined as follows:

Definition 30.6. Let f be a function. f is called *one-to-one*, if and only if $(x,y) \in f$ and $(z,y) \in f$ imply $x = z$. In other words, $f(x) = f(z)$, if and only if $x = z$.

We then have the following theorem:

Theorem 30.7. The inverse of a function f exists, if and only if f is one-to-one.

Proof. If f is one-to-one, then $(x,y) \in f$ and $(z,y) \in f$, if and only if $x = z$. But then $(y,x) \in f^c$ and $(y,z) \in f^c$, if and only if $x = z$ and by Definition 6.1 f^c is a function. Thus, by Definition 30.5, $f^c = f^{-1}$ is the inverse of f.

On the other hand, if f^{-1} exists, then $(y,x) \in f^{-1}$ and $(y,z) \in f^{-1}$, if and only if $x = z$. But $(y,x) \in f^{-1}$ and $(y,z) \in f^{-1}$, if and only if $(x,y) \in f$ and $(z,y) \in f$. Thus by Definition 30.6, f is one-to-one.

The only problem now remaining is a means for obtaining the inverse of a function when it exists. Most of the functions we have considered are those which are defined by an equation relating the first and second elements of the ordered pairs in the function. We will restrict our discussion to functions of that type hereafter.

Let f be defined by the equation $y = f(x)$ for $(x,y) \in f$. As was observed in Definition 30.5, when f^{-1} exists, $f[f^{-1}(x)] = x$ for all $x \in$ domain f^{-1}, and $f^{-1}(x) \in$ domain f. Thus, to find a formula for f^{-1}, we need only solve for $f^{-1}(x)$ in terms of x from the equation $f[f^{-1}(x)] = x$. We now give some examples to illustrate this procedure.

Example 30.8. f is defined by $f(x) = 2x + 1$. Find f^{-1}.

Solution. f is one-to-one, since for each fixed y there is only one x such that $2x + 1 = y$, namely, $x = (y - 1)/2$. To find f^{-1}, observe that $f[f^{-1}(x)] = 2f^{-1}(x) + 1 = x$. Then $2f^{-1}(x) = x - 1$, and $f^{-1}(x) = (x - 1)/2$. Thus, f^{-1} is defined by the equation $f^{-1}(x) = (x - 1)/2$. Observe that

$$f[f^{-1}(x)] = 2\left(\frac{x - 1}{2}\right) + 1 = x - 1 + 1 = x$$

and
$$f^{-1}[f(x)] = \frac{2x + 1 - 1}{2} = x.$$

Example 30.9. If f is defined by $f(x) = x^2$, $x \geq 0$, find f^{-1}.

Solution. f is one-to-one, for $x^2 = a = z^2$ with $x, z \geq 0$ imply $x = z$. Hence, f^{-1} exists. To find it, let $f[f^{-1}(x)] = [f^{-1}(x)]^2 = x$. Then $f^{-1}(x) = \sqrt{x}$, for the substitution of $f^{-1}(x)$ for x in the equation defining f implies $f^{-1}(x) \geq 0$. Note that $f[f^{-1}(x)] = [\sqrt{x}]^2 = x$ and $f^{-1}[f(x)] = \sqrt{x^2} = |x|$ but as $x \geq 0$, $|x| = x$.

Example 30.10. If f is defined by $f(x) = x^2$, $x \leq 0$, find f^{-1}.

Solution. Clearly f is one-to-one, for $x^2 = a = z^2$ with x, $z \leq 0$ imply $x = z$. Hence, f^{-1} exists and $f[f^{-1}(t)] = [f^{-1}(t)]^2 = t$. [Note $t \geq 0$ and $f^{-1}(t) \leq 0$.] Thus, $|f^{-1}(t)| = \sqrt{t}$, and as $f^{-1}(t) \leq 0$, $|f^{-1}(t)| = -f^{-1}(t)$. This gives $f^{-1}(t) = -\sqrt{t}$, or, in terms of x, $f^{-1}(x) = -\sqrt{x}$. Again, $f^{-1}[f(x)] = -\sqrt{x^2} = -|x| = -(-x) = x$, as $x \leq 0$. Also, $f[f^{-1}(x)] = (-\sqrt{x})^2 = x$.

These two examples illustrate the relation between a function and its inverse. Thus the domain of f is the range of f^{-1} and the range of f is the domain of f^{-1}. Observe that, in Example 30.10, if $g(t) = \sqrt{t}$, then $g[f(x)] = \sqrt{x^2} = |x| = -x$, as $x \leq 0$. Thus, g cannot be the inverse of f.

Another observation that could be made at this time is that the inverse is found essentially by interchanging x and y in the original equation and then solving for y in terms of x. For example, if $y = x^2$, $x \leq 0$, write $x = y^2$, $y \leq 0$. Then $|y| = \sqrt{x}$, and as $y \leq 0$, $y = -\sqrt{x}$.

Example 30.11. Let f be defined by $f(x) = a^x$, $a > 0$, $a \neq 1$. Find f^{-1}.

Solution. Clearly f is one-to-one, for $a^x = a^y$ implies $x = y$, and f^{-1} exists. Now $f[f^{-1}(x)] = a^{f^{-1}(x)} = x$. By the definition of logarithms it follows that $f^{-1}(x) = \log_a x$. Similarly, the inverse of the logarithm function to the base a is the exponential function to the base a.

EXERCISES

1. Find f^c in each of the following. Does f^{-1} exist?
   (a) $f = \{(2,1), (3,4), (1,2), (5,1)\}$
   (b) $f = \{(1,4), (2,3), (3,7), (4,9), (5,6)\}$
   (c) $f = \{(x,y) : y = |x|, x \in \{0,1,2,3,4\}\}$
   (d) $f = \{(x,y) : y = |x|, x \in \{0, \pm 1, \pm 2, \pm 3\}\}$.

Find the inverse of each of the following if it exists. If it doesn't exist show why not.

2. $f = \{(x,y) : y = x + 1\}$.
3. $f = \{(x,y) : y = 3x - 2\}$.
4. $f = \{(x,y) : y = x^3\}$.
5. $f = \{(x,y) : y = x^3 - 1\}$.

6. $f = \{(x,y) : y = x^3 - x\}$.

7. $f = \{(x,y) : y = 4x + 3\}$.

8. $f = \{(x,y) : y = x^2 - 2x\}$.

9. $f = \{(x,y) : y = x^3 + 2x^2\}$.

10. $f = \{(x,y) : y = 2^{3x}\}$.

11. $f = \{(x,y) : y = 3^{4x+3}\}$.

12. $f = \{(x,y) : y = |x|, x \leq 0\}$.

13. $f = \{(x,y) : y = \log_5 (5x + 2)\}$.

14. $f = \{(x,y) : y = \log_{10} (x^2 + 1)\}$.

15. $f = \{(x,y) : y = \log_2 (2x + 3)\}$.

16. $f = \{(x,y) : y = 2 + \log_{10} (3x - 1)\}$.

17. $f = \{(x,y) : y = 4 \cdot 5^{2x-1}\}$.

18. If $f = \{(x,y) : y = x^2 - 1\}$, show that f is not one-to-one. Find a restriction on x which will make f one-to-one and then find f^{-1}.

19. If $f = \{(x,y) : y = x^2 - 4x + 3, x \geq 2\}$, show that f is one-to-one and find f^{-1}.

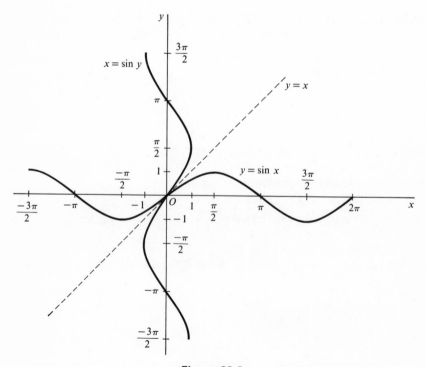

Figure 31.1

31. The Graph of f^c

From the definitions of a graph and f^c, the graph of f^c is easy to obtain from the graph of f. For the graph of f^c is the set of all points (x,y) such that the point (y,x) is in the graph of f, and this essentially interchanges the coordinate axes. Thus if f and f^c are sketched on the same coordinate axes, and the plane is folded along the line $\{(x,y) : y = x\}$, the two curves will coincide.

Example 31.1. A sketch of the graph of the sine function and its converse on the same coordinate axes is given in Figure 31.1. on page 112. The graph of the sine function is $\{(x,y) : y = \sin x\}$, and the graph of its converse is $\{(x,y) : x = \sin y\}$.

EXERCISES

Sketch the graphs of f and f^c in each of the problems of section **30**.

32. The Inverse Sine and Cosine Functions and Their Graphs

As the sine and cosine functions are periodic with period 2π, i.e., $\sin(x + n \cdot 2\pi) = \sin x$ and $\cos(x + n \cdot 2\pi) = \cos x$ for any integer n, it follows that neither the sine nor the cosine has an inverse as it stands. The converses of the sine and cosine will be called arcsine and arccosine, respectively, and given by

$$\text{arcsin} = \{(x,y) : x = \sin y\} \quad \text{or} \quad y = \text{arcsin } x \quad \text{if and only if}$$
$$x = \sin y, \quad \text{and}$$

$$\text{arccos} = \{(x,y) : x = \cos y\} \quad \text{or} \quad y = \text{arccos } x \quad \text{if and only if}$$
$$x = \cos y.$$

From this it follows, for example, that $y = \text{arcsin } \tfrac{1}{2}$ if and only if $\sin y = \tfrac{1}{2}$. Then $y = \pi/6 + 2n\pi$ and $y = 5\pi/6 + 2n\pi$ satisfy $\sin y = \tfrac{1}{2}$ for every integer n, so another way of writing this is

$$\text{arcsin } \tfrac{1}{2} = \left\{ y : y = \frac{\pi}{6} + 2n\pi \quad \text{or} \quad y = 5\pi/6 + 2n\pi, \qquad n \text{ an integer} \right\}.$$

Similarly, $y = \arccos 0$, if and only if $\cos y = 0$, and $y = \pi/2 + n\pi$ for any integer n. Thus,

$$\arccos 0 = \left\{ y : y = \frac{\pi}{2} + n\pi, \quad n \text{ any integer} \right\}.$$

This problem of not having an inverse is certainly not unique to the sine and cosine functions, nor is it impossible to solve. Recall that in Examples 30.9 and 30.10 we had the functions $\{(x,y) : y = x^2, \ x \geq 0\}$ and $\{(x,y) : y = x^2, \ x \leq 0\}$. Each of these functions has an inverse while the function $\{(x,y) : y = x^2, \ x \text{ is real}\}$ does not. Thus, we should expect that, by restricting the domains sufficiently, the functions defined by $y = \sin x$ and $y = \cos x$ will be one-to-one and hence have inverses. These will be called the "inverse sine" and "inverse cosine," respectively, and defined as follows.

Definition 32.1.   The *inverse sine function*, written $\sin^{-1}$, is defined by $\sin^{-1} = \{(x,y) : x = \sin y \text{ and } (-\pi/2) \leq y \leq (\pi/2)\}$ or $y = \sin^{-1}x$, if and only if $x = \sin y$ and $-\pi/2 \leq y \leq \pi/2$. The *inverse cosine function*, written $\cos^{-1}$, is defined by $\cos^{-1} = \{(x,y) : x = \cos y \text{ and } 0 \leq y \leq \pi\}$, or $y = \cos^{-1}x$, if and only if $x = \cos y$ and $0 \leq y \leq \pi$.

From this definition it follows immediately that $\sin (\sin^{-1} x) = x$ and $\cos (\cos^{-1} t) = t$ for all x, for which $\sin^{-1} x$ is defined and all t for which $\cos^{-1} t$ is defined. However, $\sin^{-1} (\sin x)$ is not necessarily x. For example, $\sin \pi = 0$ and $\sin^{-1} 0 = 0$. Also, $\cos 3\pi/2 = 0$, but $\cos^{-1} 0 = \pi/2$. Thus, $\sin^{-1} (\sin \pi) \neq \pi$, and $\cos^{-1} [\cos (3\pi/2)] \neq 3\pi/2$. If $(-\pi/2) \leq x \leq (\pi/2)$, then, of course, $\sin^{-1} (\sin x) = x$, and if $0 \leq x \leq \pi$, then $\cos^{-1} (\cos x) = x$.

The graphs of the inverse sine and inverse cosine functions are given in Figures 32.1 and 32.2.

It should be clear that we have been very arbitrary in our choices of interval restrictions on y in the definitions of the inverse sine and inverse cosine functions. After all, the only thing necessary for the definition of an inverse is that the function be one-to-one. However, the particular choice here becomes rather convenient as the following examples illustrate.

Example 32.2.   Evaluate $\sin (\cos^{-1}x)$ and $\cos (\sin^{-1} x)$.

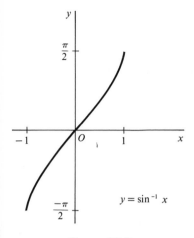

Figure 32.1   **Figure 32.2**

Solution. To compute $\sin(\cos^{-1} x)$, let $u = \cos^{-1} x$. Then $x = \cos u$ and $0 \le u \le \pi$. Thus, $x^2 = \cos^2 u = 1 - \sin^2 u$. It follows that $\sin^2 u = 1 - x^2$, and, as $0 \le u \le \pi$, $\sin u = \sqrt{1 - x^2}$. Therefore, $\sin(\cos^{-1} x) = \sqrt{1 - x^2}$.

In $\cos(\sin^{-1} x)$, let $u = \sin^{-1} x$. Then $x = \sin u$ and $(-\pi/2) \le u \le (\pi/2)$; hence $x^2 = \sin^2 u = 1 - \cos^2 u$. We then have $\cos^2 u = 1 - x^2$, and as $(-\pi/2) \le u \le (\pi/2)$, $\cos u = \sqrt{1 - x^2}$. Therefore, $\cos(\sin^{-1} x) = \sqrt{1 - x^2}$.

Example 32.3. Evaluate $\sin(\cos^{-1} x)$ and $\cos(\sin^{-1} x)$, if $y = \cos^{-1} x$ for $0 \le y < \pi/2$ or $\pi \le y \le (3\pi/2)$ and $y = \sin^{-1} x$ for $(-\pi/2) \le y \le (\pi/2)$.

Solution. Let $u = \cos^{-1} x$. Then $x = \cos u$, and either $0 \le u \le (\pi/2)$ or $\pi \le u \le (3\pi/2)$. Now $x^2 = \cos^2 u = 1 - \sin^2 u$ and $\sin^2 u = 1 - x^2$. But then $|\sin u| = \sqrt{1 - \cos^2 u} = \sqrt{1 - x^2}$ and whether $\sin(\cos^{-1} x) = \sqrt{1 - x^2}$ or $\sin(\cos^{-1} x) = -\sqrt{1 - x^2}$ depends on x. $\cos(\sin^{-1} x) = \sqrt{1 - x^2}$, just as in Example 32.2. However, by altering the range of the inverse sine function we can get a result similar to the above, namely, $\cos(\sin^{-1} x) = (\text{sign})\sqrt{1 - x^2}$.

Although these two examples do not tell the whole story, they do illustrate the importance of the choice of the range for the inverse functions.

In view of the results of Chapter 4 and the trigonometric functions related to angles, it appears advisable to relate the arcsine and arccosine to angles. This also leads to a geometrical device that is useful in evaluating certain expressions involving the inverse functions. Thus we read arcsin x as "the angle whose sine is x" and arccosine x as "the angle whose cosine is x." For example, we then have

$$\text{arcsin } \tfrac{1}{2} = \{30° + n \cdot 360°, \qquad 150° + n \cdot 360°, \qquad n \text{ an integer}\};$$

the only acute angle whose sine is $\tfrac{1}{2}$ is 30°, and so, $\sin^{-1} \tfrac{1}{2} = 30°$. We also have

$$\text{arccos } (-\tfrac{1}{2}) = \{120° + n \cdot 360°, \qquad 240° + n \cdot 360°, \qquad n \text{ an integer}\},$$

and $\cos^{-1} (-\tfrac{1}{2}) = 120°$.

The geometrical significance of the "arc functions" related to angles is obtained in the figures below. It should be obvious from Figures 32.3–32.6 that $\cos (\sin^{-1} x) = \sin (\cos^{-1} x) = \sqrt{1 - x^2}$.

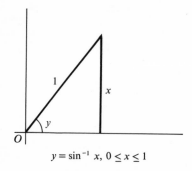

$y = \sin^{-1} x, \ 0 \le x \le 1$

Figure 32.3

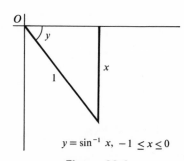

$y = \sin^{-1} x, \ -1 \le x \le 0$

Figure 32.4

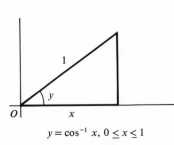

$y = \cos^{-1} x, \ 0 \le x \le 1$

Figure 32.5

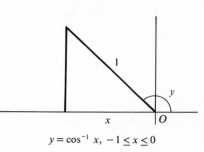

$y = \cos^{-1} x, \ -1 \le x \le 0$

Figure 32.6

Example 32.4. Evaluate $\cos (\sin^{-1} \frac{3}{5})$.

Solution. Let $y = \sin^{-1} \frac{3}{5}$. Then $\sin y = \frac{3}{5}$, y being in the first quadrant $[0 < y < (\pi/2)]$. Thus,

$$\cos y = \sqrt{1 - \sin^2 y} = \sqrt{1 - (\tfrac{3}{5})^2} = \sqrt{1 - \tfrac{9}{25}} = \sqrt{\tfrac{16}{25}} = \tfrac{4}{5}.$$

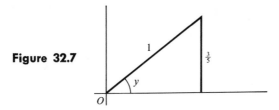

Figure 32.7

Example 32.5. Evaluate $\sin (\sin^{-1} \frac{3}{5} + \sin^{-1} \frac{2}{3})$.

Solution. Let $s = \sin^{-1} \frac{3}{5}$ and $t = \sin^{-1} \frac{2}{3}$. Then $\sin s = \frac{3}{5}$ and $\sin t = \frac{2}{3}$, as $0 < s < (\pi/2)$, $0 < t < \pi/2$. It follows that $\cos s = \frac{4}{5}$ and $\cos t = \sqrt{5}/3$. Using $\sin (s + t) = \sin s \cos t + \cos s \sin t$ and the above numbers,

$$\sin (\sin^{-1} \tfrac{3}{5} + \sin^{-1} \tfrac{2}{3}) = \tfrac{3}{5} \frac{\sqrt{5}}{3} + \tfrac{4}{5} \tfrac{2}{3} = \frac{\sqrt{5}}{5} + \frac{8}{15} = \frac{3\sqrt{5} + 8}{15}.$$

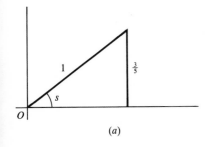
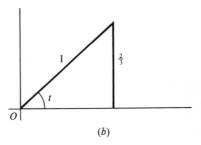

(a)   (b)

Figure 32.8

EXERCISES

1. Evaluate the following:

(a) $\cos^{-1} \left(\dfrac{\sqrt{3}}{2} \right)$ (d) $\sin^{-1} \left(\dfrac{-\sqrt{3}}{2} \right)$ (g) $\sin (\cos^{-1} \frac{1}{4})$

(b) $\cos^{-1} (-1)$ (e) $\cos^{-1} \left(\tan \dfrac{\pi}{4} \right)$ (h) $\cos (-\sin^{-1} \frac{3}{5})$

(c) $\sin^{-1} \dfrac{1}{\sqrt{2}}$ (f) $\sin^{-1} (\sec 0)$ (i) $\sin [\cos^{-1} (-\frac{1}{4})]$.

2. Evaluate the following:

(a) $\cos (\cos^{-1} \frac{5}{13} - \sin^{-1} \frac{7}{25})$

(b) $\sin (\sin^{-1} \frac{5}{13} + \sin^{-1} \frac{4}{5})$

(c) $\sin (\sin^{-1} \frac{15}{17} - \cos^{-1} \frac{7}{25})$

(d) $\sin (\sin^{-1} \frac{1}{2} - \cos^{-1} \frac{1}{3})$

(e) $\tan [\sin^{-1} (-\frac{3}{5}) - \cos^{-1} \frac{5}{13}]$

(f) $\tan [2 \sin^{-1} \frac{4}{5} + \cos^{-1} \frac{12}{13}]$

(g) $\sin [2 \sin^{-1} \frac{4}{5} - \cos^{-1} \frac{1}{3}]$

(h) $\cos (2 \sin^{-1} x)$.

3. Show that $\sin^{-1} x = \cos^{-1} \sqrt{1 - x^2}$ for $0 \le x \le \dfrac{\pi}{2}$.

4. Find and graph the inverse of the function defined by $y = \cos 2x$, $0 \le x \le \dfrac{\pi}{2}$.

33. The Other Inverse Trigonometric Functions

In the previous section, it was observed that the sine and cosine functions do not have inverses, unless suitable restrictions are made on their domains. As the other trigonometric functions are also periodic, the same is true of them. Also, the identity $\sin^2 t + \cos^2 t = 1$ affected the choice of the interval for a domain. In a similar manner, the identity $\tan^2 t + 1 = \sec^2 t$ to some extent determines the choices of intervals for the inverse tangent and inverse secant functions. The inverse cotangent and inverse cosecant will be left as exercises. The arctangent and arcsecant are defined first and then the inverse functions are defined.

Definition 33.1. The arctangent and arcsecant are defined by

$$\text{arctangent} = \{(x,y) : (y,x) \in \text{tangent}\}$$

$$= \{(x,y) : x = \tan y\},$$

and $\qquad \text{arcsecant} = \{(x,y) : x = \sec y\}.$

Thus, $y = \arctan x$, if and only if $x = \tan y$, and $y = \text{arcsec } x$, if and only if $x = \sec y$.

As in the case of the arcsine and arccosine, arctan x and arcsec x are read, respectively, "a number whose tangent is x or an angle whose tangent is x," and "a number whose secant is x or an angle whose secant is x." For example, $\arctan 0 = \{n\pi : n \text{ an integer}\} = \{n \ 180° : n \text{ an integer}\}$, $\arctan 1 = \{(\pi/4) + n\pi : n \text{ an integer}\} = \{45° + n \cdot 180° : n \text{ an integer}\}$, $\text{arcsec } 1 = \{n \cdot 2\pi : n \text{ an integer}\} = \{n \cdot 360° : n \text{ an integer}\}$, and $\text{arcsec } 2 = \{(\pi/3) + n \cdot 2\pi, (5\pi/3) + n \cdot 2\pi : n \text{ an integer}\} = \{60° + n \cdot 360°, 300° + n \cdot 360° : n \text{ an integer}\}$.

Definition 33.2. The inverse tangent and inverse secant, written $\tan^{-1}$ and $\sec^{-1}$, respectively, are defined by

$$\tan^{-1} = \left\{ (x,y) : x = \tan y \quad \text{and} \quad -\frac{\pi}{2} < y < \frac{\pi}{2} \right\},$$

and $\quad \sec^{-1} = \left\{ (x,y) : x = \sec y \quad \text{and} \quad 0 \leq y < \frac{\pi}{2} \quad \text{or} \quad \pi \leq y < \frac{3\pi}{2} \right\}.$

Thus, $y = \tan^{-1} x$, if and only if $x = \tan y$ and $-\pi/2 < y < \pi/2$, and $y = \sec^{-1} x$ if and only if $x = \sec y$ and $0 \leq y < \pi/2$ or $\pi \leq y < 3\pi/2$.

The graphs of the inverse tangent and inverse secant are given in Figures 33.1 and 33.2.

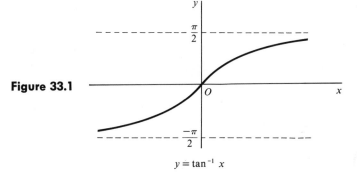

Figure 33.1

$$y = \tan^{-1} x$$

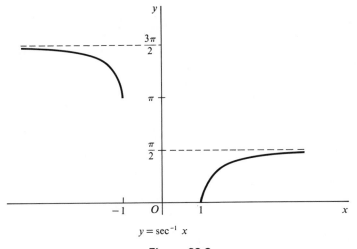

$$y = \sec^{-1} x$$

Figure 33.2

Example 33.3. Show that $\tan^{-1} \frac{1}{2} + \tan^{-1} \frac{1}{3} = \pi/4$.

Solution. Let $\tan^{-1} \frac{1}{2} = s$ and $\tan^{-1} \frac{1}{3} = t$. Then $\tan s = \frac{1}{2}$ and $\tan t = \frac{1}{3}$ with $0 < s < (\pi/2)$ and $0 < t < (\pi/2)$. Now

$$\tan(s+t) = \frac{\tan s + \tan t}{1 - \tan s \tan t}$$

$$= \frac{\frac{1}{2} + \frac{1}{3}}{1 - \frac{1}{2} \cdot \frac{1}{3}} = \frac{\frac{5}{6}}{\frac{5}{6}} = 1.$$

Thus, as $0 < s + t < \pi$, it follows that $s + t = \pi/4$.

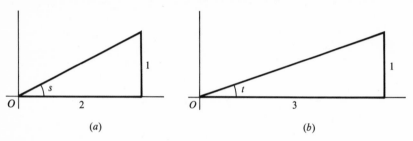

(a)   (b)

Figure 33.3

EXERCISES

1. Evaluate the following:

(a) $\tan^{-1} \dfrac{1}{\sqrt{3}}$ (d) $\sec^{-1}\left(\sin \dfrac{\pi}{2}\right)$ (g) $\sec^{-1}(-2)$

(b) $\tan^{-1}(-\sqrt{3})$ (e) $\tan(\sec^{-1} 2)$ (h) $\sec(\cos^{-1} \frac{1}{2})$

(c) $\tan(\sin^{-1} 0)$ (f) $\sin(\tan^{-1} 2)$ (i) $\cos[\tan^{-1}(1)]$.

2. Show that:

(a) $\sin^{-1} \dfrac{2u}{1+u^2} = 2\tan^{-1} u$ for $-1 \le u \le 1$

(b) $\cos^{-1} \dfrac{1-u^2}{1+u^2} = 2\tan^{-1} u$ for $u \ge 0$.

3. Verify the following:

(a) $\tan^{-1} 4 + \tan^{-1} \frac{5}{3} = \dfrac{3\pi}{4}$

(b) $\tan^{-1} 1 + \tan^{-1} \frac{1}{2} = \tan^{-1} 3$

(c) $\sin^{-1}(-\frac{4}{5}) + \cos^{-1} \dfrac{2}{\sqrt{5}} = \tan^{-1}(-\frac{1}{2})$

(d) $\tan^{-1}\frac{1}{2} + \tan^{-1}\frac{1}{5} + \tan^{-1}\frac{1}{8} = \frac{\pi}{4}$

(e) $\tan^{-1}\frac{32}{43} - \tan^{-1}\frac{1}{4} = \cos^{-1}\frac{12}{13}$

(f) $2\tan^{-1}\frac{1}{3} + \tan^{-1}\frac{1}{7} = \cos^{-1}\dfrac{5}{\sqrt{34}} + \sin^{-1}\dfrac{1}{\sqrt{17}}$

(g) $\tan^{-1}\frac{1}{2} + \tan^{-1}\frac{1}{3} = \frac{\pi}{4}$

(h) $\sin^{-1}\frac{4}{5} + \tan^{-1}\frac{3}{4} = \frac{\pi}{2}$

(i) $\tan^{-1}\frac{4}{3} - \tan^{-1}\frac{1}{7} = \frac{\pi}{4}$

(j) $2\tan^{-1}\frac{1}{3} + \tan^{-1}\frac{1}{7} = \frac{\pi}{4}$

(k) $\tan^{-1}x + \tan^{-1}\dfrac{1}{x} = \dfrac{\pi}{2}, \ x > 0.$

34. Trigonometric Equations

In Chapter 2, it was observed that there are two kinds of equations, identities and conditional equations. Nearly all the equations considered so far have been identities. To conclude the chapter we will discuss the solutions of trigonometric equations. Several examples are given below to illustrate some methods of solution.

Example 34.1. Solve the equation, $2\sin^2 t + \sin t - 1 = 0$.

Solution. As

$$2\sin^2 t + \sin t - 1 = (2\sin t - 1)(\sin t + 1),$$

it follows that $2\sin^2 t + \sin t - 1 = 0$, if and only if either $2\sin t - 1 = 0$ or $\sin t + 1 = 0$. Hence, $\sin t = \frac{1}{2}$ or $\sin t = -1$. There are an infinite number of values t for which $\sin t = \frac{1}{2}$, namely, $t = (\pi/6) + 2n\pi$ or $t = (5\pi/6) + 2n\pi$ for n any integer. There are also an infinite number of solutions of the equation $\sin t = -1$, $t = (3\pi/2) + 2n\pi$ for n any integer. If one is only interested in the solutions for $0 \le t < 2\pi$, then $t = \pi/6, 5\pi/6, 3\pi/2$ are the appropriate values.

Example 34.2. Solve the equation $\sin t + \cos t = 1$ for $0 \le t < 2\pi$.

Solution. The equation $\sin t + \cos t = 1$ is equivalent to the equation $\sin t = 1 - \cos t$. Then squaring both sides gives

$$\sin^2 t = (1 - \cos t)^2 = 1 - 2\cos t + \cos^2 t.$$

Thus,

$$1 - \cos^2 t = 1 - 2 \cos t + \cos^2 t,$$

or

$$2 \cos^2 t - 2 \cos t = 0,$$

and

$$\cos t(1 - \cos t) = 0.$$

The solutions of this derived equation are $t = 0, \pi/2, 3\pi/2$. However, $3\pi/2$ is not a solution of the equation $\sin t + \cos t = 1$. Thus the only solutions are $t = 0, \pi/2$. Note that an extraneous root was introduced when the two sides were squared.

Example 34.3.   Solve the equation, $3 \sin^2 t = 1, 0 \le t < 2\pi$.

Solution. $3 \sin^2 t = 1$, if and only if $\sin^2 t = \frac{1}{3}$. Then extracting square roots gives $|\sin t| = 1/\sqrt{3}$. Now $\sin t = 1/\sqrt{3}$, when $0 \le t \le \pi$ and $\sin t = -1/3$, when $\pi < t < 2\pi$. The equation, $\sin t = 1/\sqrt{3}$ has two solutions, which may be written as $t = \sin^{-1} (1/\sqrt{3})$ and $t = \pi - \sin^{-1} (1/\sqrt{3})$, and $\sin t = -1/\sqrt{3}$ has two solutions which may be written as $t = \pi + \sin^{-1} (1/\sqrt{3})$ and $t = 2\pi - \sin^{-1} (1/\sqrt{3})$. For the student interested in finding an explicit representation for $\sin^{-1} (1/\sqrt{3})$, from Table I, it follows that $\sin^{-1} (1/\sqrt{3})$ is about $35\frac{1}{3}°$ or 0.614.

Example 34.4.   Solve the equation, $\sin 2t + \sin t = 0$ for $0 \le t < 2\pi$.

Solution.   As $\sin 2t = 2 \sin t \cos t$, the equation may be written as

$$2 \sin t \cos t + \sin t = 0.$$

Then

$$2 \sin t \cos t + \sin t = \sin t(2 \cos t + 1) = 0,$$

if and only if $\sin t = 0$ or $2 \cos t + 1 = 0$. Thus

$$t = 0, \pi, \quad \text{and} \quad t = \frac{2\pi}{3}, \frac{4\pi}{3}$$

are solutions of the respective equations, hence also of the original equation.

Example 34.5.   Solve the equation $\sin^2 t + 2 \cos t = 1$.

Solution.   As $1 - \sin^2 t = \cos^2 t$, the above equation is equivalent to

$$2 \cos t = \cos^2 t.$$

Then

$$\cos^2 t - 2 \cos t = \cos t \,(\cos t - 2) = 0.$$

The only solutions of this equation arise from $\cos t = 0$, as $\cos t \neq 2$. Thus, $t = \pi/2$ and $3\pi/2$ are the only solutions of the original equation.

Although there are no general methods for solving trigonometric equations, the above examples illustrate three important procedures that may be used in solving most trigonometric equations. They are factoring, in Examples 34.1, 34.3, and 34.4, writing the equations in terms of a single function in Example 34.5, and squaring both members in Example 34.2. In many cases identities are useful in working with equations as is seen in Examples 34.4 and 34.5. Another thing that should be observed is that the functions in the equation should all have the same argument. For example, in the equation, $\sin x + \cos 4x = 0$, $\cos 4x$ should be reduced to an expression involving only $\sin x$ and $\cos x$.

EXERCISES

Solve the following equations for $0 \leq x < 2\pi$.

1. $\sin 2x = \sqrt{3} \sin x$.

2. $\tan^2 x = 1$.

3. $\tan^2 x = 3$.

4. $2 \sin^2 x - 5 \sin x - 3 = 0$.

5. $\cos x + \cos 2x = 0$.

6. $2 \cos x + \sec x = 3$.

7. $\sec x - 1 = \tan x$.

8. $\sin 3x + \sin x = 0$.

9. $\sin \dfrac{x}{2} + \cos x = 1$.

10. $\cos 2x + \cos 3x = 0$. *Hint:* $\cos s + \cos t = 2 \cos \dfrac{s+t}{2} \cos \dfrac{s-t}{2}$.

11. $\cos 3x + \cos 4x = 0$.

12. $\sin 2x + \sin 3x = 0$. *Hint:* $\sin s + \sin t = 2 \sin \dfrac{s+t}{2} \cos \dfrac{s-t}{2}$.

13. $\sin 3x + \sin 4x = 0$.

14. $\sin 2x + \sin 4x = 2 \sin 3x$.

15. $\cos 5x + \cos x = 2 \cos 2x$.

16. $\sin x + \sin 3x = \cos x + \cos 3x$.

6

COMPLEX NUMBERS

35. Definitions and Fundamental Concepts

Recall that in Chapter 1, section **8**, complex numbers were introduced as ordered pairs of real numbers. Thus, the set of complex numbers is the set $\{(a,b) : a,b \text{ are real}\}$ with addition and multiplication defined by

$$(a,b) + (c,d) = (a + c, b + d),$$

and

$$(a,b)(c,d) = (ac - bd, ad + bc).$$

Two complex numbers, (a,b) and (c,d), were said to be equal,

$$(a,b) = (c,d), \quad \text{if and only if} \quad a = c \quad \text{and} \quad b = d.$$

From the above definition of addition, it follows that any complex number (a,b) may be written as

$$(a,b) = (a,0) + (0,b).$$

Furthermore, for any complex number (a,b),

$$(a,b)(1,0) = (a,b),$$

so $(1,0)$ has the same properties under multiplication as "1" does in the real numbers. Thus, we will simply write "1" instead of $(1,0)$. In fact, if we define $(a,0) = a(1,0)$, then we may write "a," instead of $(a,0)$.

The complex number $(0,1)$ has the property that $(0,1)^2 = (-1,0) = -1$. Thus, if we define $i = (0,1)$ and $(0,b) = b(0,1)$, then the complex number $(0,b)$ is written as bi.

On the basis of the above discussion we now write the complex number (a,b) as

$$(a,b) = (a,0) + (0,b) = a(1,0) + b(0,1)$$
$$= a + bi.$$

The number a is called the real part and b the imaginary part of the complex number. The sum of $a + bi$ and $c + di$ is now

$$(a + bi) + (c + di) = (a + c) + (b + d)i.$$

The product of $a + bi$ and $c + di$ is

$$(a + bi)(c + di) = (ac - bd) + (ad + bc)i.$$

Note that the addition and multiplication of complex numbers are the same as that of binomials of the form $a + bx$ and $c + dx$.

To illustrate these new "definitions" of addition and multiplication, we have for example:

$$(2 + 3i) + (1 - 5i) : 2 + 3i$$
$$\frac{1 - 5i}{3 - 2i}$$

$$(1 - i)(3 + 2i) : 1 - i$$
$$\frac{3 + 2i}{3 - 3i}$$
$$\frac{2i - 2i^2}{(3 + 2) + (2 - 3)i},$$

or $(1 - i)(3 + 2i) = 5 - i,$ as $i^2 = -1.$

Subtraction of complex numbers is done in the obvious manner, i.e.,
$(a + bi) - (c + di) = (a + bi) + [(-c) + (-d)i] = (a - c) + (b - d)i$.
From Property (M4′) of section **8**, as

$$(a,b)\left[\frac{a}{a^2 + b^2}, \frac{b}{a^2 + b^2}\right] = (1,0) \quad \text{when} \quad a^2 + b^2 \neq 0,$$

it follows that

$$(a + bi)\left[\frac{a}{a^2 + b^2} - \frac{b}{a^2 + b^2} i\right] = 1.$$

Thus, the reciprocal of $a + bi$, $\dfrac{1}{a + bi}$, is defined by

$$\frac{1}{a + bi} = \frac{a}{a^2 + b^2} - \frac{b}{a^2 + b^2} i,$$

for $a + bi \neq 0$. Using this and the definition of multiplication, it follows that

$$\frac{a + bi}{c + di} = (a + bi) \frac{1}{c + di}$$

$$= (a + bi)\left[\frac{c}{c^2 + d^2} - \frac{d}{c^2 + d^2} i\right]$$

$$= \frac{ac + bd}{c^2 + d^2} + \frac{bc - ad}{c^2 + d^2} i.$$

Thus, for example,

$$\frac{1 + i}{2 - i} = (1 + i) \frac{1}{2 - i} = (1 + i)\left[\frac{2}{2^2 + 1^2} - \frac{(-1)}{2^2 + 1^2} i\right]$$

$$= (1 + i)(\tfrac{2}{5} + \tfrac{1}{5} i)$$

$$= \frac{2 - 1}{5} + \frac{2 + 1}{5} i = \tfrac{1}{5} + \tfrac{3}{5} i.$$

EXERCISES

1. If $z_1 = a_1 + b_1 i$ and $z_2 = a_2 + b_2 i$ are complex numbers, show that $z_1 + z_2 = z_2 + z_1$.

2. Evaluate the following:

(a) $(1 + 2i)(1 - i)$ (b) $3(2 - i)$

(c) $2i(1 + i)$ (g) $\dfrac{3}{2 - i}$

(d) $(3 + 2i)(4 + i)$ (h) $\dfrac{2i}{1 + i}$

(e) $(1 - i)(1 - i)$ (i) $\dfrac{3 + 2i}{4 + i}$

(f) $\dfrac{1 + 2i}{1 - i}$ (j) $\dfrac{1 + i}{2 + i}$.

3. (a) Prove $i^3 = -i$, $i^4 = 1$

   (b) Prove $i^{4k} = 1$, $i^{4k-1} = -i$, $i^{4k-2} = -1$ and $i^{4k-3} = i$ for every positive integer k.

4. Solve the following equations using the definition of equality.

   (a) $2x - yi = 4 + 3i$

   (b) $(x + yi)(1 + 2i) = 3 - 4i$

   (c) $(x - yi)(4 - i) = 1 + 4i$.

36. Rectangular Form and Geometric Significance of Complex Numbers

The definition of complex numbers as ordered pairs of real numbers allows for an immediate geometric interpretation of complex numbers. For, given the complex number (a,b), there is exactly one point in the plane having coordinates (a,b). Since the complex number $(a,b) = a + bi$, $a + bi$ is called the rectangular form of the complex number.

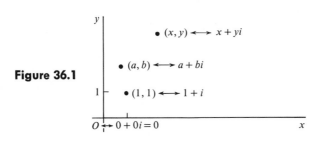

Figure 36.1

Observe from Figure 36.1 that $0 + 0i$ is the origin. Furthermore, the real numbers have y coordinate zero, and the x coordinate of the pure imaginary numbers is zero. Thus the x axis is called the real axis, the y axis is called the imaginary axis, and the coordinate plane is the complex plane.

Associated with every point in the plane is the real number which represents its distance from the origin. Thus if (a,b) is any point, its distance from the origin is $\sqrt{a^2 + b^2}$. In the same manner, then, with every complex number $a + bi$ is associated a real number called its modulus or absolute value, indicated by $|a + bi|$, and defined by $|a + bi| = \sqrt{a^2 + b^2}$. Thus, the absolute value of a complex number is the square root of the sum of the squares of its real and imaginary parts.

The definition of addition of two complex numbers (a,b) and (c,d) as given in section **8** leads to an interesting geometric interpretation. For $(a,b) + (c,d) = (a + c, b + d)$ (See Figure 36.2 below), and this sum is

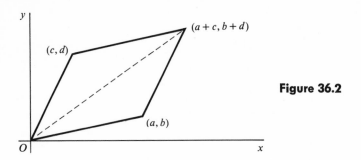

Figure 36.2

the end point of the diagonal of the parallelogram having the segments from the origin to the respective points (a,b) and (c,d) as sides. If the complex numbers are proportional, then they and their sum lie on the same line passing through the origin. The student should use the distance formula to verify that the quadrilateral thus formed in Figure 36.2 is a parallelogram.

One other algebraic concept having geometric significance is that of the conjugate. If $a + bi$ is a complex number, its conjugate, $\overline{a + bi}$, is defined by $\overline{a + bi} = a - bi$. Geometrically, the complex numbers $a + bi$ and $a - bi$ are symmetric with respect to the x axis.

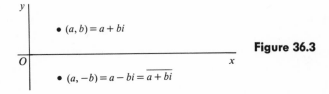

Figure 36.3

The complex conjugate is extremely useful. For example,

$$(a + bi)\,\overline{(a + bi)} = (a + bi)\,(a - bi) = a^2 + b^2$$
$$= |a + bi|^2.$$

Thus, it follows that

$$|a + bi| = \sqrt{(a + bi)\,\overline{(a + bi)}}.$$

Also, $(a + bi) + \overline{(a + bi)} = (a + bi) + (a - bi) = 2a,$

or the sum of a complex number and its conjugate is twice the real part of the complex number. A number of properties involving conjugates can be proved quite easily using such definitions as sums, products, equality, and others. We will give the proof of one of these to indicate a possible procedure and list some others as exercises.

Example 36.1.   Prove

$$\overline{(a + bi) + (c + di)} = \overline{(a + bi)} + \overline{(c + di)}.$$

Proof. $\overline{(a + bi) + (c + di)} = \overline{(a + c) + (b + d)i}$

$$= (a + c) - (b + d)i$$

$$\overline{(a + bi)} + \overline{(c + di)} = (a - bi) + (c - di)$$

$$= (a + c) - (b + d)i.$$

Then from the definition of equality of complex numbers, as the respective real and imaginary parts of the two numbers are equal, it follows that

$$\overline{(a + bi) + (c + di)} = \overline{(a + bi)} + \overline{(c + di)}.$$

EXERCISES

1. Prove the following:
   (a) $\overline{(a + bi)\,(c + di)} = \overline{(a + bi)}\,\overline{(c + di)}$
   (b) $\overline{\overline{(a + bi)}} = a + bi$

(c) $\overline{a + bi} = a + bi$ if and only if $b = 0$

(d) $|\overline{a + bi}| = |a + bi|$

(e) $\overline{\left(\dfrac{a + bi}{c + di}\right)} = \dfrac{\overline{a + bi}}{\overline{c + di}}$

(f) $\dfrac{a + bi}{c + di} = \dfrac{(a + bi(\overline{c + di})}{(c + di)(\overline{c + di})}$

(g) $|(a + bi)(c + di)| = |a + bi||c + di|$.

2. Plot the following pairs of complex numbers in a coordinate plane and find their sum.

   (a) $1 + 2i$, $3 - 4i$

   (b) $2 + 3i$, $1 + i$

   (c) $3 + i$, $1 - 2i$.

3. Use the distance formula to verify that the quadrilateral in Figure 36.2 is a parallelogram.

4. Find:

   (a) $|2 + i|$

   (b) $|3 - 2i|$

   (c) $|4 + 5i|$.

37. Polar Form of Complex Numbers

The geometry of the plane and polar coordinates in the plane lead to some interesting applications of trigonometry to complex numbers. As we have seen, there is a relation between complex numbers and points in the plane, which gives the amplitude of the complex number as the distance from the corresponding point to the origin. Associated with every point (x,y) in rectangular coordinates is a point $[r,\theta]$ in polar coordinates, where $x = r \cos \theta$ and $y = r \sin \theta$. θ can be any angle from the positive x axis to the segment between (x,y) and the origin. (See Figure 37.1 below.) Thus $r = \sqrt{x^2 + y^2}$ and $\tan \theta = y/x$.

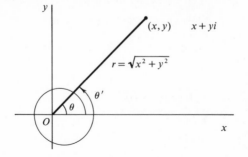

Figure 37.1

With this as a basis, if $x + yi$ is a complex number in rectangular form, its polar form is given by $r(\cos \theta + i \sin \theta)$. That is, $x + yi = r \cos \theta + ir \sin \theta = r(\cos \theta + i \sin \theta)$, with $r^2 = \sqrt{x^2 + y^2}$ and $\tan \theta = y/x$. The angle θ is called the argument of the complex number.

It is convenient to be able to change from rectangular form to polar form and vice versa. Keep in mind, though, that the polar form for a particular complex number is not unique. For example, $2(\cos 37° + i \sin 37°) = 2(\cos 397° + i \sin 397°) = \cdots$.

Example 37.1. Write $5(\cos 30° + i \sin 30°)$ in rectangular form.

Solution. As $\cos 30° = \sqrt{3}/2$ and $\sin 30° = \frac{1}{2}$,

$$5(\cos 30° + i \sin 30°) = 5\left(\frac{\sqrt{3}}{2} + \frac{1}{2} i\right)$$

$$= \frac{5\sqrt{3}}{2} + \frac{5}{2} i.$$

Example 37.2. Write $1 - i$ in polar form.

Solution. $r^2 = 1^2 + 1^2 = 2$, and so $r = \sqrt{2}$. $\tan \theta = y/x = -1/1$ has as a solution $\theta = 7\pi/4$. Hence,

$$1 - i = \sqrt{2}\left(\cos \frac{7\pi}{4} + i \sin \frac{7\pi}{4}\right).$$

Another choice for θ might have been $\theta = 15\pi/4$, for example.

The change from rectangular to polar form simplifies the multiplication and division of complex numbers, and also the raising of complex numbers to powers and the extraction of roots. The multiplication, division and powers are taken up now, and the extraction of roots is considered in the next section.

Theorem 37.3. Let $z_1 = r(\cos \theta + i \sin \theta)$ and $z_2 = s(\cos \phi + i \sin \phi)$ be complex numbers in polar form. Then

(a) $z_1 \cdot z_2 = rs \, [\cos (\theta + \phi) + i \sin (\theta + \phi)]$,

and   (b) $\dfrac{z_1}{z_2} = \dfrac{r}{s} \, [\cos (\theta - \phi) + i \sin (\theta - \phi)]$.

Proof of (a).

$$r(\cos\theta + i\sin\theta)\cdot s(\cos\phi + i\sin\phi)$$

$$= rs[(\cos\theta\cos\phi - \sin\theta\sin\phi)$$

$$+ i(\sin\theta\cos\phi + \cos\theta\sin\phi)]$$

$$= rs[\cos(\theta + \phi) + i\sin(\theta + \phi)].$$

The proof of (b) is left as an exercise.

Stated in words, part (a) says that the argument of the product is the sum of the arguments and the amplitude of the product is the product of the amplitudes. Similarly, in part (b), we have the argument of the quotient is the difference of the arguments, and the amplitude of the quotient is the quotient of the amplitudes.

As a special case of (a), observe that if $z = r(\cos\theta + i\sin\theta)$, then $z^2 = r^2(\cos 2\theta + i\sin 2\theta)$. The generalization of this is called DeMoivre's Theorem which we will now prove.

Theorem 37.4 (DeMoivre's Theorem). If $z = r(\cos\theta + i\sin\theta)$ is any complex number in polar form and n is any positive integer, then $z^n = r^n(\cos n\theta + i\sin n\theta)$.

Proof. The proof is by mathematical induction. First observe that the theorem is true for $n = 1$ and has been verified above for $n = 2$. Suppose now it is true for $n = k$, i.e.,

$$z^k = r^k(\cos k\theta + i\sin k\theta).$$

Then

$$z^{k+1} = z\cdot z^k = [r(\cos\theta + i\sin\theta)][r^k(\cos k\theta + i\sin k\theta)]$$

$$= r^{k+1}[\cos(\theta + k\theta) + i\sin(\theta + k\theta)] \text{ by Theorem 37.3 (a)}$$

$$= r^{k+1}[\cos(k + 1)\theta + i\sin(k + 1)\theta].$$

Hence, it is true for $n = k + 1$ when it is true for $n = k$, and as it is true for $n = 1$, it is true for every positive integer n.

Example 37.5. Find $(1 + i)^6$.

Solution. In polar form the complex number $1 + i$ takes the form $1 + i = \sqrt{2}[\cos(\pi/4) + i \sin(\pi/4)]$. Hence,

$$(1 + i)^6 = \left[\sqrt{2}\left(\cos\frac{\pi}{4} + i\sin\frac{\pi}{4}\right)\right]^6$$

$$= (\sqrt{2})^6\left[\cos 6\left(\frac{\pi}{4}\right) + i\sin 6\left(\frac{\pi}{4}\right)\right]$$

$$= 8\left(\cos\frac{3\pi}{2} + i\sin\frac{3\pi}{2}\right)$$

$$= -8i, \text{ in rectangular form.}$$

EXERCISES

1. Prove Theorem 37.3 (b).

2. Compute the following and write in rectangular form:
   (a) $5(\cos 70° + i\sin 70°)2(\cos 50° + i\sin 50°)$
   (b) $3(\cos 185° + i\sin 185°)4(\cos 40° + i\sin 40°)$
   (c) $\dfrac{8(\cos 335° + i\sin 335°)}{3(\cos 110° + i\sin 110°)}$
   (d) $\dfrac{5(\cos 40° + i\sin 40°)}{2(\cos 100° + i\sin 100°)}.$

3. Use DeMoivre's Theorem to compute the following and write in rectangular form.
   (a) $(1 - i)^{10}$ (d) $(2 + \sqrt{3}i)^6$
   (b) $(-1 + i)^4$ (e) $(\sqrt{3} - i)^3$
   (c) $(-1 - i)^5$ (f) $(-1 - \sqrt{3}i)^8.$

4. Let $\omega = \cos\dfrac{2\pi}{3} + i\sin\dfrac{2\pi}{3}.$ Show that $\omega^2 + \omega + 1 = 0$.

5. Extend DeMoivre's Theorem to include all negative integers as well. Thus, if $r \neq 0$, then $[r(\cos\theta + i\sin\theta)]^{-n} = r^{-n}[\cos(-n\theta) + i\sin(-n\theta)] = r^{-n}(\cos n\theta - i\sin n\theta)$.

38. Extracting nth Roots of Complex Numbers

Closely associated with powers of complex numbers are the roots of complex numbers. As for real numbers, we say that, if n is a positive integer and z and w are complex numbers such that $w^n = z$, then w is called an *nth root* of z. To find the *n*th roots of a complex number we proceed as follows.

Let $z = r(\cos \theta + i \sin \theta)$ and $w = s(\cos \phi + i \sin \phi)$ be complex numbers such that w is an nth root of z. Then

$$w^n = s^n(\cos n\phi + i \sin n\phi) = r(\cos \theta + i \sin \theta).$$

By equality of complex numbers it follows that $s^n = r$, $\cos n\phi = \cos \theta$, and $\sin n\phi = \sin \theta$. From these equations it follows that $s = r^{1/n}$ and $n\phi = \theta + k2\pi$ for some integer k. This leads to the conclusion that $\phi = \dfrac{\theta + k2\pi}{n}$, or that an nth root of z has the form

$$w_k = z^{1/n} = r^{1/n}\left(\cos \frac{\theta + k2\pi}{n} + i \sin \frac{\theta + k2\pi}{n}\right),$$

where k is some integer.

We now observe several things about the numbers w_k. First of all, $w_k = w_{k+n}$ for all integers k. For

$$w_{k+n} = r^{1/n}\left[\cos \frac{\theta + (k + n)2\pi}{n} + i \sin \frac{\theta + (k + n)2\pi}{n}\right]$$

$$= r^{1/n}\left[\cos \frac{\theta + k2\pi + n2\pi}{n} + i \sin \frac{\theta + k2\pi + n2\pi}{n}\right]$$

$$= r^{1/n}\left[\cos\left(\frac{\theta + k2\pi}{n} + 2\pi\right) + i \sin\left(\frac{\theta + k2\pi}{n} + 2\pi\right)\right]$$

$$= r^{1/n}\left[\cos \frac{\theta + k2\pi}{n} + i \sin \frac{\theta + k2\pi}{n}\right]$$

$$= w_k.$$

In particular, if $0 \le k < n$, then $w_k = w_{k+n}$, so $w_0 = w_n$, $w_1 = w_{n+1}$, etc.

Second, $w_k \ne w_m$ for $0 \le k < n$, $0 \le m < n$ and $k \ne m$. For the equations

$$\cos \frac{\theta + k2\pi}{n} = \cos \frac{\theta + m2\pi}{n}$$

and

$$\sin \frac{\theta + k2\pi}{n} = \sin \frac{\theta + m2\pi}{n}$$

imply that k and m differ by multiples of n.

From these two observations we may now say that the n distinct nth roots of $z = r(\cos \theta + i \sin \theta)$ are

$$w_k = r^{1/n}\left(\cos \frac{\theta + k2\pi}{n} + i \sin \frac{\theta + k2\pi}{n}\right), \ k = 0, 1, 2, \cdots, n - 1.$$

If the complex numbers w_k are plotted in polar coordinates in the plane,

$$w_k = \left[r^{1/n}, \frac{\theta}{n} + \frac{k2\pi}{n}\right].$$

Thus, the points w_k are located on a circle of radius $r^{1/n}$ and spaced equally about the circumference. The angle between any two consecutive points is $2\pi/n$. See Figure 38.1.

Example 38.1. Find the three cube roots of i.

Solution. In polar form the complex number i is written

$$i = 1\left(\cos \frac{\pi}{2} + i \sin \frac{\pi}{2}\right).$$

Then the cube roots are

$$w_k = 1\left\{\cos \frac{\frac{\pi}{2} + k2\pi}{3} + i \sin \frac{\frac{\pi}{2} + k2\pi}{3}\right\}, \ k = 0, 1, 2.$$

The polar and rectangular forms of the roots are thus given by

$$w_0 = \cos \frac{\pi}{6} + i \sin \frac{\pi}{6} = \frac{\sqrt{3}}{2} + \tfrac{1}{2} i$$

$$w_1 = \cos \frac{\frac{5\pi}{2}}{3} + i \sin \frac{\frac{5\pi}{2}}{3} = \cos \frac{5\pi}{6} + i \sin \frac{5\pi}{6} = \frac{-\sqrt{3}}{2} + \tfrac{1}{2} i$$

$$w_2 = \cos \frac{\frac{9\pi}{2}}{3} + i \sin \frac{\frac{9\pi}{2}}{3} = \cos \frac{3\pi}{2} + i \sin \frac{3\pi}{2} = -i.$$

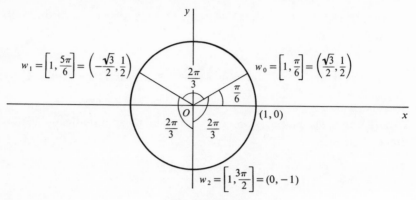

Figure 38.1

The three cube roots of i are illustrated in Figure 38.1.

Example 38.2. Find the fourth roots of -1.

Solution. The polar form for the complex number -1 is

$$-1 = 1(\cos \pi + i \sin \pi).$$

Then the fourth roots of -1 are

$$w_k = \cos \frac{\pi + k2\pi}{4} + i \sin \frac{\pi + k2\pi}{4}, \; k = 0, 1, 2, 3.$$

Then the polar and rectangular forms of the roots are given as follows:

$$w_0 = \cos \frac{\pi}{4} + i \sin \frac{\pi}{4} = \frac{1}{\sqrt{2}} + \frac{1}{\sqrt{2}} i$$

$$w_1 = \cos \frac{\pi + 2\pi}{4} + i \sin \frac{\pi + 2\pi}{4} = \cos \frac{3\pi}{4} + i \sin \frac{3\pi}{4}$$

$$= \frac{-1}{\sqrt{2}} + \frac{1}{\sqrt{2}} i$$

$$w_2 = \cos \frac{\pi + 4\pi}{4} + i \sin \frac{\pi + 4\pi}{4} = \cos \frac{5\pi}{4} + i \sin \frac{5\pi}{4}$$

$$= \frac{-1}{\sqrt{2}} - \frac{1}{\sqrt{2}} i$$

$$w_3 = \cos \frac{\pi + 6\pi}{4} + i \sin \frac{\pi + 6\pi}{4} = \cos \frac{7\pi}{4} + i \sin \frac{7\pi}{4}$$

$$= \frac{1}{\sqrt{2}} - \frac{1}{\sqrt{2}} i.$$

EXERCISES

Find the indicated roots and express them in both polar and rectangular form, unless indicated otherwise.

1. The square roots of $2 - 2i\sqrt{3}$.

2. The fourth roots of $-4 - 4i\sqrt{3}$.

3. The cube roots of $-8 + 8i$ in polar form.

4. The sixth roots of $-i$.

5. The sixth roots of 8.

6. The cube roots of $8i$.

7. The fourth roots of i in polar form.

8. The fifth roots of -32 in polar form.

9. The cube roots of $2 - 2i\sqrt{3}$ in polar form.

10. If ω is a complex eighth root of 1, show that $\bar{\omega}$ must also be an eighth root of 1.

11. If ω is any complex fifth root of 1, show that $\omega^4 + \omega^3 + \omega^2 + \omega + 1 = 0$. Show also that the other four roots have the form, or are equal to ω^2, ω^3, ω^4, ω^5. Hence, conclude that it is possible to find all the fifth roots by finding one and then using DeMoivre's Theorem.

12. Apply the procedure outlined in Problem 11 to find the seventh roots of 1 in polar form.

13. Show that the procedure of Problem 11 fails to yield all the sixth roots of 1 for two values of ω. That is, find a sixth root, ω_1, of 1 such that $\omega_1^2 = 1$ and a sixth root, ω_2, of 1 such that $\omega_2^3 = 1$.

APPENDIX A

LOGARITHMS AND EXPONENTS

Definition 1.   If a is any real number and n is a positive integer, then

$$a^n = a \cdot a \cdots a, \qquad n \text{ factors } a.$$

n is called the exponent, a the base, and a^n an exponential or the nth power of a.

Then the following rules of exponents can be shown to hold:

(1) $a^n a^m = a^{n+m}$.

(2) $(ab)^n = a^n b^n$.

(3) $(a^n)^m = a^{nm}$.

(4) $\left(\dfrac{a}{b}\right)^n = \dfrac{a^n}{b^n}, \qquad b \neq 0.$

Definition 2. If $a \neq 0$ is any real number and n is a positive integer, then

$$a^{-n} = \frac{1}{a^n}.$$

Using this definition it is also easy to show that

$$a^{-n}a^{-m} = a^{-n-m} = a^{-(n+m)}.$$

We would like to have all the above rules of exponents apply when m and n are arbitrary integers, positive, negative, or zero. In particular, $a^n a^{-n} = a^n(1/a^n) = 1$, so we define $a^0 = 1$ for $a \neq 0$. Then $a^n a^{-n} = a^{n-n} = a^0 = 1$. Thus, with this criterion and the above definitions, it follows that rules (1)–(4) above hold for all integers and appropriate numbers a and b, and in addition we have the following:

(5) $\dfrac{a^m}{a^n} = a^{m-n}, \qquad m > n.$

(6) $\dfrac{a^m}{a^n} = \dfrac{1}{a^{n-m}}, \qquad n > m.$

Definition 3. If a is any real number, n a positive integer, and b is a real number such that $b^n = a$, then b is called an nth root of a, written $b = \sqrt[n]{a}$. If $a < 0$, then in order for $b^n = a$, n must be an odd integer.

As there are two real nth roots of any positive number with n an even integer, it is necessary to distinguish between them in some way. Thus, if $a > 0$, then by $\sqrt[n]{a}$ we will mean the *positive* number b such that $b^n = a$. If $a < 0$ and n an odd integer, then by $\sqrt[n]{a}$, we will mean the *negative* number b such that $b^n = a$. For example, then, $\sqrt{4} = 2$ not ± 2, $\sqrt[3]{8} = 2$, $\sqrt[4]{625} = 5$, and $\sqrt[3]{-27} = -3$.

By definition it follows that $(\sqrt[n]{a})^n = a$. Also if $a^{1/n}$ is to be defined in such a way that $(a^{1/n})^n = a^{n/n} = a$, we need only let $a^{1/n} = \sqrt[n]{a}$. Using this as a definition, we have the following properties:

(7) $a^{m/n} = (a^m)^{1/n} = (a^{1/n})^m = \sqrt[n]{a^m} = (\sqrt[n]{a})^m, \qquad a > 0$

and m, n integers, $n > 0$.

(8) $(a^{1/n})^{1/m} = a^{1/nm}.$

It is possible to extend the notion of exponents to include such expressions as $2^{\sqrt{3}}$ and $(\sqrt{3})^{\sqrt{2}}$. Such a development is impossible here, so now assume that if $a > 0$, $a \neq 1$, and x an arbitrary real number, then a^x is defined so that the following rules hold.

(E1) $a^0 = 1$.

(E2) $a^x a^y = a^{x+y}$.

(E3) $(a^x)^y = a^{xy}$.

(E4) $(ab)^x = a^x b^x$.

(E5) $\left(\dfrac{a}{b}\right)^x = \dfrac{a^x}{b^x}$, $\quad a \neq 0$.

(E6) $\dfrac{a^x}{a^y} = a^{x-y}$, $\quad x > y$.

(E7) $\dfrac{a^x}{a^y} = \dfrac{1}{a^{y-x}}$, $\quad y > x$.

With the above ideas as a basis we now define the exponential functions.

Definition 4. If $a > 0$, $a \neq 1$, then the exponential function to the base a, written $\exp_a$, is defined by

$$\exp_a = \{(x,y) : y = a^x\}.$$

Thus if $(x,y) \in \exp_a$, then we may also write $y = \exp_a(x)$.

As an example, if $a = 10$, then $\exp_{10} = \{(x,y) : y = 10^x\}$. Then $(0,1) \in \exp_{10}$, $(1,10) \in \exp_{10}$, $(-1,\frac{1}{10}) \in \exp_{10}$, $(10,10^{10}) \in \exp_{10}$, etc.

Logarithms are defined in terms of exponentials. Thus we have:

Definition 5. If $a > 0$, $a \neq 1$, then the logarithm to the base a of x equals y, written $y = \log_a x$, if and only if $x = a^y$. The logarithm function, $\log_a$, is defined by $\log_a = \{(x,y) : y = \log_a x\}$. As $a^y > 0$ for all y, it follows that $\log_a x$ is not defined for $x \leq 0$.

Using the definitions of logarithms and exponentials, and the properties of exponentials, the following properties of logarithms may be derived.

(L1) $\log_a MN = \log_a M + \log_a N$.

(L2) $\log_a \dfrac{M}{N} = \log_a M - \log_a N$.

(L3) $\log_a M^p = p \log_a M$.

(L4) $\log_a a = 1$.

(L5) $\log_a 1 = 0$.

Our principal use of logarithms here is in computation, and as the decimal system uses the number base 10, that will also be the logarithm base used in the tables. Thus we will write $\log x$ for $\log_{10} x$.

In scientific notation if $M > 0$ is a real number, then there is an integer c and a number m, $1 \le m < 10$, such that

$$M = 10^c m.$$

Then using (L1) and (L3),

$$\log M = \log 10^c m = \log 10^c + \log m$$
$$= c + \log m.$$

As $1 \le m < 10$, $0 \le \log m < 1$, assuming that $x < y$, if and only if $10^x < 10^y$. The integer c is called the characteristic of the logarithm and m, the mantissa. The big advantage, then, in using the base 10, is that we need only have tables of logarithms for numbers from 1 to 10.

The characteristic of the logarithm of a number can be found using either of the following equivalent methods: (1) Write the number in scientific notation and the characteristic is the power to which 10 is raised. (2) If the number is greater than or equal to 1, the characteristic is 1 less than the number of digits to the left of the decimal point; if the number is less than 1, the characteristic is negative and 1 less than the number of zeros to the right of the decimal and to the left of the first nonzero digit. Thus the characteristic of $\log 298.3$ is 2, and that of $\log 0.00037$ is -4.

The table of logarithms found here, Table III (page 165), is a four-place table and allows the finding of logarithms of the numbers between 1 and 10 which are given to the nearest hundredth. Linear interpolation can be used with this table to find logarithms of numbers between 1 and 10, expressed to the nearest thousandth, but this method is not developed here. The student should refer to a good college algebra text to find an explanation of interpolation.

We will now give examples to illustrate the use of logarithms.

Example 1. Compute

$$\frac{(1.23)\,(23.8)\sqrt{29}}{(312)\,(0.115)}.$$

We first write

$$\log \frac{(1.23)\,(23.8)\sqrt{29}}{(312)\,(0.115)}$$

$$= \log (1.23)\,(23.8)\sqrt{29} - \log (312)\,(0.115)$$

$$= \log 1.23 + \log 23.8 + \log 29^{\frac{1}{2}} - [\log 312 + \log 0.115]$$

$$= \log 1.23 + \log 23.8 + \tfrac{1}{2} \log 29 - [\log 312 + \log 0.115]$$

From Table III,

$$
\begin{aligned}
\log 1.23 &= & 0 + 0.0899 \\
\log 23.8 &= & 1 + 0.3766 \\
\tfrac{1}{2} \log 29 &= & \tfrac{1}{2}(1 + 0.4624) \\
&= & 0 + 0.7312 \\
\hline
\log \text{numerator} &= & 2 + 0.1977 \\
\log 312 &= & 2 + 0.4942 \\
\log 0.115 &= & -1 + 0.0607 \\
&= & 9 + 0.0607 - 10 \\
\hline
\log \text{denominator} &= & 11 + 0.5549 - 10 \\
&= & 1 + 0.5549 \\
\log \text{numerator} &= & 2 + 0.1977 \\
- \log \text{denominator} &= & 1 + 0.5549 \\
\hline
\log \text{fraction} &= & 0 + 0.6428
\end{aligned}
$$

Then from Table III (page 165), we find that log 4.39 = 0.6415 and log 4.4 = 0.6425. Thus the answer is approximately 4.39.

Example 2. Compute $(-0.00324)(7.21)$.

Even though the number is negative, we can use logarithms to compute $(0.00324)(7.21)$ and then insert the minus sign again when we have finished:

$$\log (0.00324)(7.21) = \log 0.00324 + \log 7.21.$$

Then from Table III(page 165),

$$
\begin{aligned}
\log 0.00324 &= -3 + 0.5105 \\
&= 7 + 0.5105 - 10 \\
\log 7.21 &= 0 + 0.8579 \\
\hline
\log \text{product} &= 8 + 0.3684 - 10.
\end{aligned}
$$

Then the number whose mantissa is closest to 0.3684 is 2.34, and the characteristic is -2. Thus $(-0.00324)(7.21) = -0.0234$, approximately.

In both of these examples we have written negative characteristics as $k - 10$ for a positive integer $k < 10$. That is, $-1 = 9 - 10$ and $-3 = 7 - 10$. This was done because the mantissas are positive numbers. The characteristic only determines where the decimal point should be placed.

EXERCISES

1. Find the characteristic of the logarithm of each of the numbers.

(a) 34.5	(e) log 2790	(i) log 386,000
(b) 273.5	(f) 0.000159	(k) 3.25
(c) 5937.4	(g) 2 log 1005	(l) $38 \cdot 10^{-5}$
(d) 0.147	(h) $\sin \dfrac{\pi}{6}$	(m) $295 \cdot 10^{-9}$.

2. Use Table III (page 165) to find the logarithms of the following.

(a) 345	(c) 5,930	(e) 0.000159	(g) 0.00105
(b) 27.3	(d) 1.47	(f) 2.79	(h) 368,000.

3. If $\log 2 = 0.3010$ and $\log 3 = 0.4771$, find:

(a) log 6	(d) log 5.4
(b) log 8	(e) log 0.5
(c) log 0.024	(f) log 72.

4. Use Table III (page 165) to find the approximate value of x in each of the following.

(a) $5^x = 10$

(b) $10^{5x} = 32$

(c) $x = 10^{2.3}$

(d) $10^x = 60$

(e) $10^{2x+1} = 85$

(f) $10^{3x-1} = 20$.

5. Use logarithms to compute the following.

(a) $\sqrt[3]{\dfrac{(235)\,(129)}{547}}$

(b) $\dfrac{\sqrt[4]{87}}{\sqrt{23}}$

(c) $\dfrac{(-89.7)\,(432)\,(5.71)}{6,340}$

(d) $\dfrac{(0.0313)\,(-4.27)\,(3.85)}{(0.00121)\,(15.8)\,(35.7)}$

(e) $\dfrac{\sqrt[3]{35}}{\sqrt[5]{27}}$

(f) $(1.59)^4\,(27.8)^{\frac{1}{3}}\,(9.83)^2$

(g) $\dfrac{(2.34)^2\,(19.8)^3}{\sqrt{765}}$

(h) $\dfrac{15.3\sqrt{91}}{\sqrt[3]{84}}$.

6. Solve the following equations.

(a) $\log_2 x = 4$

(b) $\log_3 x = \frac{1}{2}$

(c) $\log_x 9 = 3$

(d) $\log_3 9 = x$

(e) $\log_2 x + \log_2(4 - x) = 2$

(f) $\log_{10} (x + 3) - \log_{10} x = 1$.

LAW OF TANGENTS AND
TANGENTS OF HALF-ANGLES

The Law of Tangents may be used to solve an oblique triangle given two sides and the included angle. This is used in place of the Law of Cosines if one wants to use logarithms in the solution of the triangle. The Law of Tangents is derived below.

From the Law of Sines,

$$\frac{\sin \alpha}{a} = \frac{\sin \beta}{b}, \quad \text{or} \quad \frac{a}{b} = \frac{\sin \alpha}{\sin \beta}.$$

Then adding 1 to both sides gives

$$\frac{a}{b} + 1 = \frac{a+b}{b} = \frac{\sin \alpha}{\sin \beta} + 1 = \frac{\sin \alpha + \sin \beta}{\sin \beta}.$$

Subtracting 1 from both sides gives

$$\frac{a}{b} - 1 = \frac{a-b}{b} = \frac{\sin \alpha}{\sin \beta} - 1 = \frac{\sin \alpha - \sin \beta}{\sin \beta}.$$

Dividing the first expression by the second yields

$$\frac{a+b}{a-b} = \frac{\sin \alpha + \sin \beta}{\sin \alpha - \sin \beta}.$$

Figure B.1

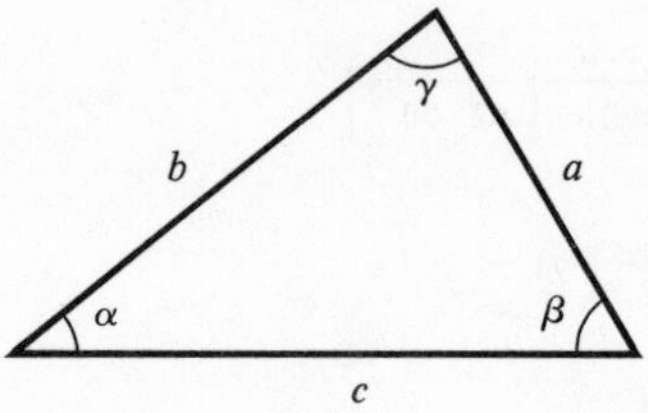

As $\sin x + \sin y = 2 \sin \frac{1}{2}(x + y) \cos \frac{1}{2}(x - y)$, and $\sin x - \sin y = 2 \sin \frac{1}{2}(x - y) \cos \frac{1}{2}(x + y)$, it follows that

(1)
$$\frac{a+b}{a-b} = \frac{\tan \frac{1}{2}(\alpha + \beta)}{\tan \frac{1}{2}(\alpha - \beta)}.$$

In a similar manner we can derive

(2)
$$\frac{a+c}{a-c} = \frac{\tan \frac{1}{2}(\alpha + \gamma)}{\tan \frac{1}{2}(\alpha - \gamma)},$$

and (3)
$$\frac{b+c}{b-c} = \frac{\tan \frac{1}{2}(\beta + \gamma)}{\tan \frac{1}{2}(\beta - \gamma)}.$$

In using the Law of Tangents, if $a < b$, then the formula

$$\frac{b+a}{b-a} = \frac{\tan \frac{1}{2}(\beta + \alpha)}{\tan \frac{1}{2}(\beta - \alpha)}$$

should be used. If a, b, and γ are given, then $a + b, a - b$, and $\frac{1}{2}(\alpha + \beta)$ $= 90° - \frac{\gamma}{2}$ are easy to find, and from these we obtain $\frac{1}{2}(\alpha - \beta)$. Then, as $\frac{1}{2}(\alpha + \beta) + \frac{1}{2}(\alpha - \beta) = \alpha$ and $\frac{1}{2}(\alpha + \beta) - \frac{1}{2}(\alpha - \beta) = \beta$, α and β are easily found. Then the Law of Sines is used to find c.

The tangents of half-angle formulas are used to find the angles given the three sides of a triangle. As with the Law of Tangents, this method can be used in place of the Law of Cosines and is more convenient if it is desirable to use logarithms. The derivation is given below.

Figure B.2

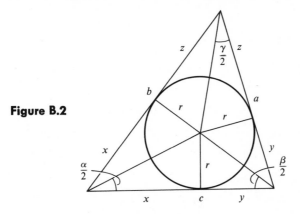

From Problem 3, section **27**, the radius of the inscribed circle is

$$r = \sqrt{\frac{(s-a)(s-b)(s-c)}{s}},$$

where $s = \frac{1}{2}(a+b+c)$. Then $x+y+z = s$. But $x+y = c$, $x+z = b$, and $y+z = a$, so $s-z = c$, $s-y = b$, and $s-x = a$. From Figure B.2,

$$\tan\frac{\alpha}{2} = \frac{r}{x},$$

$$\tan\frac{\beta}{2} = \frac{r}{y},$$

and
$$\tan\frac{\gamma}{2} = \frac{r}{z}.$$

Then using the value for r given above and the values of x, y, and z, it follows that

$$\tan\frac{\alpha}{2} = \sqrt{\frac{(s-b)(s-c)}{s(s-a)}}$$

(4)
$$\tan\frac{\beta}{2} = \sqrt{\frac{(s-a)(s-c)}{s(s-b)}}$$

$$\tan\frac{\gamma}{2} = \sqrt{\frac{(s-a)(s-b)}{s(s-c)}}.$$

These are the half-angle formulas.

The examples below illustrate the use of the Law of Tangents and half-angle formulas.

Example 1.   Solve the triangle with $a = 12.5$, $b = 15.4$ and $\gamma = 54°$.

Solution.   The Law of Tangents may be used. As $b > a$, we will use the formula

$$\frac{b + a}{b - a} = \frac{\tan \frac{1}{2}(\beta + \alpha)}{\tan \frac{1}{2}(\beta - \alpha)}.$$

Then $b + a = 27.9$, $b - a = 2.9$, and $\alpha + \beta = 180° - \gamma = 126°$, or $\frac{1}{2}(\beta + \alpha) = 63°$. Thus,

$$\frac{27.9}{2.9} = \frac{\tan 63°}{\tan \frac{1}{2}(\beta - \alpha)},$$

or
$$\tan \frac{1}{2}(\beta - \alpha) = \frac{2.9 \tan 63°}{27.9}.$$

Using logarithms we write

$$\log \tan \frac{1}{2}(\beta - \alpha) = \log \frac{2.9 \tan 63°}{27.9}$$

$$= \log 2.9 + \log \tan 63° - \log 27.9.$$

From Table III, $\log 2.9 = 0.4624$ and $\log 27.9 = 1.4456$, and from Table IV, $\log \tan 63° = 0.2928$. Thus

$$\log \tan \frac{1}{2}(\beta - \alpha) = 0.4624 + 0.2928 - 1.4456$$

$$= 10.7552 - 1.4456 - 10$$

$$= 9.3096 - 10.$$

From Table IV (page 167), $\frac{1}{2}(\beta - \alpha) = 12°$, to the nearest degree. Thus,

$$\beta - \alpha = 24°,$$

$$\beta + \alpha = 126°,$$

and $\beta = 75°$, $\alpha = 51°$.

Having found α and β, we apply the Law of Sines to find c.

$$\frac{\sin \alpha}{a} = \frac{\sin \gamma}{c},$$

or

$$c = \frac{a \sin \gamma}{\sin \alpha} = \frac{12.5 \sin 54°}{\sin 51°}$$

$$= 13, \text{ approximately.}$$

Example 2. Solve the triangle with $a = 13.3$, $b = 16.4$, and $c = 15.1$.

Solution. The half-angle formulas may be used to solve this triangle. Using formula (4) above, we have

$$\tan \frac{\alpha}{2} = \sqrt{\frac{(s - b)(s - c)}{s(s - a)}},$$

$$\tan \frac{\beta}{2} = \sqrt{\frac{(s - a)(s - c)}{s(s - b)}},$$

$$s = \tfrac{1}{2}(a + b + c) = \tfrac{1}{2}(13.3 + 16.4 + 15.1) = 22.4$$

$$s - a = 9.1, \ s - b = 6.0, \ s - c = 7.3,$$

$$\tan \frac{\alpha}{2} = \sqrt{\frac{(6.0)(7.3)}{(22.4)(9.1)}}.$$

Then

$$\log \tan \frac{\alpha}{2} = \tfrac{1}{2} \log \frac{(6.0)(7.3)}{(22.4)(9.1)}$$

$$= \tfrac{1}{2}[\log (6.0)(7.3) - \log (22.4)(9.1)]$$

$\log 6.0 = 0.7782$	$\log 22.4 = 1.3502$
$\log 7.3 = 0.8633$	$\log 9.1 \ \ = 0.9590$
$\log$ numerator $= 1.6415$	$\log$ denominator $= 2.4092$

$$\log \text{ quotient} = 19.2323 - 20.$$

Thus,

$$\log \tan \frac{\alpha}{2} = \tfrac{1}{2}(19.2323 - 20) = 9.6162 - 10.$$

From Table IV (page 167), $\alpha/2 = 22.5°$, to the nearest half-degree, and $\alpha = 45°$, to the nearest degree.

Next we have

$$\tan \frac{\beta}{2} = \sqrt{\frac{(9.1)\,(7.3)}{(22.4)\,(6.0)}}.$$

Then,

$$\log \tan \frac{\beta}{2} = \tfrac{1}{2} \log \frac{(9.1)\,(7.3)}{(22.4)\,(6.0)}$$

$$= \tfrac{1}{2}\,[\log(9.1)\,(7.3) - \log(22.4)\,(6.0)]$$

$\log 9.1 = 0.9590$	$\log 22.4 = 1.3502$
$\log 7.3 = 0.8633$	$\log \;\; 6.0 = 0.7782$
$\log$ numerator $= 1.8223$	$\log$ denominator $= 2.1284$

$$\log \text{ quotient } = 19.6939 - 20.$$

Thus,

$$\log \tan \frac{\beta}{2} = \tfrac{1}{2}(19.6939 - 20) = 9.8464 - 10.$$

From Table IV (page 167), $\beta/2 = 35°$, to the nearest half-degree, and so $\beta = 70°$, to the nearest degree. Then $\gamma = 180° - (\alpha + \beta) = 180° - (70° + 45°) = 65°$.

EXERCISES

Solve the following triangles using the Law of Tangents or half-angle formulas. For other problems using these formulas, refer to section **26.**

1. $a = 23.4$, $b = 19.5$, $\gamma = 44°$.

2. $b = 453$, $c = 501$, $\alpha = 50°$.

3. $a = 0.125$, $c = 0.153$, $\beta = 40°$.

4. $a = 7.83$, $b = 6.94$, $c = 7.39$.

5. $a = 27.4$, $b = 33.7$, $c = 36.5$.

6. $a = 125$, $b = 138$, $c = 151$.

TABLES

The following tables are from Plane Trigonometry *by Nathan O. Niles,*
copyright 1959, published by John Wiley and Sons.
Reproduced by permission of the publisher.

TABLE I

Four-Place Values of Trigonometric Functions
Angle θ in Degrees and Radians

Angle θ Degrees	Angle θ Radians	sin θ	csc θ	tan θ	cot θ	sec θ	cos θ		
0° 00′	.0000	.0000	No value	.0000	No value	1.000	1.0000	1.5708	90° 00′
10	029	029	343.8	029	343.8	000	000	679	50
20	058	058	171.9	058	171.9	000	000	650	40
30	.0087	.0087	114.6	.0087	114.6	1.000	1.0000	1.5621	30
40	116	116	85.95	116	85.94	000	.9999	592	20
50	145	145	68.76	145	68.75	000	999	563	10
1° 00′	.0175	.0175	57.30	.0175	57.29	1.000	.9998	1.5533	89° 00′
10	204	204	49.11	204	49.10	000	998	504	50
20	233	233	42.98	233	42.96	000	997	475	40
30	.0262	.0262	38.20	.0262	38.19	1.000	.9997	1.5446	30
40	291	291	34.38	291	34.37	000	996	417	20
50	320	320	31.26	320	31.24	001	995	388	10
2° 00′	.0349	.0349	28.65	.0349	28.64	1.001	.9994	1.5359	88° 00′
10	378	378	26.45	378	26.43	001	993	330	50
20	407	407	24.56	407	24.54	001	992	301	40
30	.0436	.0436	22.93	.0437	22.90	1.001	.9990	1.5272	30
40	465	465	21.49	466	21.47	001	989	243	20
50	495	494	20.23	495	20.21	001	988	213	10
3° 00′	.0524	.0523	19.11	.0524	19.08	1.001	.9986	1.5184	87° 00′
10	553	552	18.10	553	18.07	002	985	155	50
20	582	581	17.20	582	17.17	002	983	126	40
30	:0611	.0610	16.38	.0612	16.35	1.002	.9981	1.5097	30
40	640	640	15.64	641	15.60	002	980	068	20
50	669	669	14.96	670	14.92	002	978	039	10
4° 00′	.0698	.0698	14.34	.0699	14.30	1.002	.9976	1.5010	86° 00′
10	727	727	13.76	729	13.73	003	974	981	50
20	756	756	13.23	758	13.20	003	971	952	40
30	.0785	.0785	12.75	0787	12.71	1.003	.9969	1.4923	30
40	814	814	12.29	816	12.25	003	967	893	20
50	844	843	11.87	846	11.83	004	964	864	10
5° 00′	.0873	.0872	11.47	.0875	11.43	1.004	.9962	1.4835	85° 00′
10	902	901	11.10	904	11.06	004	959	806	50
20	931	929	10.76	934	10.71	004	957	777	40
30	.0960	.0958	10.43	.0963	10.39	1.005	.9954	1.4748	30
40	989	987	10.13	992	10.08	005	951	719	20
50	.1018	.1016	9.839	.1022	9.788	005	948	690	10
6° 00′	.1047	.1045	9.567	.1051	9.514	1.006	.9945	1.4661	84° 00′
		cos θ	sec θ	cot θ	tan θ	csc θ	sin θ	Radians	Degrees
								Angle θ	

TABLE I—*continued*

Angle θ Degrees	Radians	sin θ	csc θ	tan θ	cot θ	sec θ	cos θ		
6° 00′	.1047	.1045	9.567	.1051	9.514	1.006	.9945	1.4661	84° 00′
10	076	074	9.309	080	9.255	006	942	632	50
20	105	103	9.065	110	9.010	006	939	603	40
30	.1134	.1132	8.834	.1139	8.777	1.006	.9936	1.4573	30
40	164	161	8.614	169	8.556	007	932	544	20
50	193	190	8.405	198	8.345	007	929	515	10
7° 00′	.1222	.1219	8.206	.1228	8.144	1.008	.9925	1.4486	83° 00′
10	251	248	8.016	257	7.953	008	922	457	50
20	280	276	7.834	287	7.770	008	918	428	40
30	.1309	.1305	7.661	.1317	7.596	1.009	.9914	1.4399	30
40	338	334	7.496	346	7.429	009	911	370	20
50	367	363	7.337	376	7.269	009	907	341	10
8° 00′	.1396	.1392	7.185	.1405	7.115	1.010	.9903	1.4312	82° 00′
10	425	421	7.040	435	6.968	010	899	283	50
20	454	449	6.900	465	6.827	011	894	254	40
30	.1484	.1478	6.765	.1495	6.691	1.011	.9890	1.4224	30
40	513	507	6.636	524	6.561	012	886	195	20
50	542	536	6.512	554	6.435	012	881	166	10
9° 00′	.1571	.1564	6.392	.1584	6.314	1.012	.9877	1.4137	81° 00′
10	600	593	277	614	197	013	872	108	50
20	629	622	166	644	084	013	868	079	40
30	.1658	.1650	6.059	.1673	5.976	1.014	.9863	1.4050	30
40	687	679	5.955	703	871	014	858	1.4021	20
50	716	708	855	733	769	015	853	992	10
10° 00′	.1745	.1736	5.759	.1763	5.671	1.015	.9848	1.3963	80° 00′
10	774	765	665	793	576	016	843	934	50
20	804	794	575	823	485	016	838	904	40
30	.1833	.1822	5.487	.1853	5.396	1.017	.9833	1.3875	30
40	862	851	403	883	309	018	827	846	20
50	891	880	320	914	226	018	822	817	10
11° 00′	.1920	.1908	5.241	.1944	5.145	1.019	.9816	1.3788	79° 00′
10	949	937	164	974	066	019	811	759	50
20	978	965	089	.2004	4.989	020	805	730	40
30	.2007	.1994	5.016	.2035	4.915	1.020	.9799	1.3701	30
40	036	.2022	4.945	065	843	021	793	672	20
50	065	051	876	095	773	022	787	643	10
12° 00′	.2094	.2079	4.810	.2126	4.705	1.022	.9781	1.3614	78° 00′
10	123	108	745	156	638	023	775	584	50
20	153	136	682	186	574	024	769	555	40
30	.2182	.2164	4.620	.2217	4.511	1.024	.9763	1.3526	30
40	211	193	560	247	449	025	757	497	20
50	240	221	502	278	390	026	750	468	10
13° 00′	.2269	.2250	4.445	.2309	4.331	1.026	.9744	1.3439	77° 00′
		cos θ	sec θ	cot θ	tan θ	csc θ	sin θ	Radians	Degrees
								Angle θ	

154

TABLE I—*continued*

| Angle θ | | | | | | | | | |
Degrees	Radians	sin θ	csc θ	tan θ	cot θ	sec θ	cos θ		
13° 00′	.2269	.2250	4.445	.2309	4.331	1.026	.9744	1.3439	77° 00′
10	298	278	390	339	275	027	737	410	50
20	327	306	336	370	219	028	730	381	40
30	.2356	.2334	4.284	.2401	4.165	1.028	.9724	1.3352	30
40	385	363	232	432	113	029	717	323	20
50	414	391	182	462	061	030	710	294	10
14° 00′	.2443	.2419	4.134	.2493	4.011	1.031	.9703	1.3265	76° 00′
10	473	447	086	524	3.962	031	696	235	50
20	502	476	039	555	914	032	689	206	40
30	.2531	.2504	3.994	.2586	3.867	1.033	.9681	1.3177	30
40	560	532	950	617	821	034	674	148	20
50	589	560	906	648	776	034	667	119	10
15° 00′	.2618	.2588	3.864	.2679	3.732	1.035	.9659	1.3090	75° 00′
10	647	616	822	711	689	036	652	061	50
20	676	644	782	742	647	037	644	032	40
30	.2705	.2672	3.742	.2773	3.606	1.038	.9636	1.3003	30
40	734	700	703	805	566	039	628	974	20
50	763	728	665	836	526	039	621	945	10
16° 00′	.2793	.2756	3.628	.2867	3.487	1.040	.9613	1.2915	74° 00′
10	822	784	592	899	450	041	605	886	50
20	851	812	556	931	412	042	596	857	40
30	.2880	.2840	3.521	.2962	3.376	1.043	.9588	1.2828	30
40	909	868	487	994	340	044	580	799	20
50	938	896	453	.3026	305	045	572	770	10
17° 00′	.2967	.2924	3.420	.3057	3.271	1.046	.9563	1.2741	73° 00′
10	996	952	388	089	237	047	555	712	50
20	.3025	979	357	121	204	048	546	683	40
30	.3054	.3007	3.326	.3153	3.172	1.048	.9537	1.2654	30
40	083	035	295	185	140	049	528	625	20
50	113	062	265	217	108	050	520	595	10
18° 00′	.3142	.3090	3.236	.3249	3.078	1.051	.9511	1.2566	72° 00′
10	171	118	207	281	047	052	502	537	50
20	200	145	179	314	018	053	492	508	40
30	.3229	.3173	3.152	.3346	2.989	1.054	.9483	1.2479	30
40	258	201	124	378	960	056	474	450	20
50	287	228	098	411	932	057	465	421	10
19° 00′	.3316	.3256	3.072	.3443	2.904	1.058	.9455	1.2392	71° 00′
10	345	283	046	476	877	059	446	363	50
20	374	311	021	508	850	060	436	334	40
30	.3403	.3338	2.996	.3541	2.824	1.061	.9426	1.2305	30
40	432	365	971	574	798	062	417	275	20
50	462	393	947	607	773	063	407	246	10
20° 00′	.3491	.3420	2.924	.3640	2.747	1.064	.9397	1.2217	70° 00′
		cos θ	sec θ	cot θ	tan θ	csc θ	sin θ	Radians	Degrees
								Angle θ	

TABLE I—*continued*

Angle θ		sin θ	csc θ	tan θ	cot θ	sec θ	cos θ		
Degrees	**Radians**								
20° 00′	.3491	.3420	2.924	.3640	2.747	1.064	.9397	1.2217	**70° 00′**
10	520	448	901	673	723	065	387	188	50
20	549	475	878	706	699	066	377	159	40
30	.3578	.3502	2.855	.3739	2.675	1.068	.9367	1.2130	30
40	607	529	833	772	651	069	356	101	20
50	636	557	812	805	628	070	346	072	10
21° 00′	.3665	.3584	2.790	.3839	2.605	1.071	.9336	1.2043	**69° 00′**
10	694	611	769	872	583	072	325	1.2014	50
20	723	638	749	906	560	074	315	985	40
30	.3752	.3665	2.729	.3939	2.539	1.075	.9304	1.1956	30
40	782	692	709	973	517	076	293	926	20
50	811	719	689	.4006	496	077	283	897	10
22° 00′	.3840	.3746	2.669	.4040	2.475	1.079	.9272	1.1868	**68° 00′**
10	869	773	650	074	455	080	261	839	50
20	898	800	632	108	434	081	250	810	40
30	.3927	.3827	2.613	.4142	2.414	1.082	.9239	1.1781	30
40	956	854	595	176	394	084	228	752	20
50	985	881	577	210	375	085	216	723	10
23° 00′	.4014	.3907	2.559	.4245	2.356	1.086	.9205	1.1694	**67° 00′**
10	043	934	542	279	337	088	194	665	50
20	072	961	525	314	318	089	182	636	40
30	.4102	.3987	2.508	.4348	2.300	1.090	.9171	1.1606	30
40	131	.4014	491	383	282	092	159	577	20
50	160	041	475	417	264	093	147	548	10
24° 00′	.4189	.4067	2.459	.4452	2.246	1.095	.9135	1.1519	**66° 00′**
10	218	094	443	487	229	096	124	490	50
20	247	120	427	522	211	097	112	461	40
30	.4276	.4147	2.411	.4557	2.194	1.099	.9100	1.1432	30
40	305	173	396	592	177	100	088	403	20
50	334	200	381	628	161	102	075	374	10
25° 00′	.4363	.4226	2.366	.4663	2.145	1.103	.9063	1.1345	**65° 00′**
10	392	253	352	699	128	105	051	316	50
20	422	279	337	734	112	106	038	286	40
30	.4451	.4305	2.323	.4770	2.097	1.108	.9026	1.1257	30
40	480	331	309	806	081	109	013	228	20
50	509	358	295	841	066	111	001	199	10
26° 00′	.4538	.4384	2.281	.4877	2.050	1.113	.8988	1.1170	**64° 00′**
10	567	410	268	913	035	114	975	141	50
20	596	436	254	950	020	116	962	112	40
30	.4625	.4462	2.241	.4986	2.006	1.117	.8949	1.1083	30
40	654	488	228	.5022	1.991	119	936	054	20
50	683	514	215	059	977	121	923	1.1025	10
27° 00′	.4712	.4540	2.203	.5095	1.963	1.122	.8910	1.0996	**63° 00′**
		cos θ	**sec θ**	**cot θ**	**tan θ**	**csc θ**	**sin θ**	**Radians**	**Degrees**
								Angle θ	

TABLE I—*continued*

Angle θ Degrees	Radians	sin θ	csc θ	tan θ	cot θ	sec θ	cos θ		
27° 00′	.4712	.4540	2.203	.5095	1.963	1.122	.8910	1.0996	63° 00′
10	741	566	190	132	949	124	897	966	50
20	771	592	178	169	935	126	884	937	40
30	.4800	.4617	2.166	.5206	1.921	1.127	.8870	1.0908	30
40	829	643	154	243	907	129	857	879	20
50	858	669	142	280	894	131	843	850	10
28° 00′	.4887	.4695	2.130	.5317	1.881	1.133	.8829	1.0821	62° 00′
10	916	720	118	354	868	134	816	792	50
20	945	746	107	392	855	136	802	763	40
30	.4974	.4772	2.096	.5430	1.842	1.138	.8788	1.0734	30
40	.5003	797	085	467	829	140	774	705	20
50	032	823	074	505	816	142	760	676	10
29° 00′	.5061	.4848	2.063	.5543	1.804	1.143	.8746	1.0647	61° 00′
10	091	874	052	581	792	145	732	617	50
20	120	899	041	619	780	147	718	588	40
30	.5149	.4924	2.031	.5658	1.767	1.149	.8704	1.0559	30
40	178	950	020	696	756	151	689	530	20
50	207	975	010	735	744	153	675	501	10
30° 00′	.5236	.5000	2.000	.5774	1.732	1.155	.8660	1.0472	60° 00′
10	265	025	1.990	812	720	157	646	443	50
20	294	050	980	851	709	159	631	414	40
30	.5323	.5075	1.970	.5890	1.698	1.161	.8616	1.0385	30
40	352	100	961	930	686	163	601	356	20
50	381	125	951	969	675	165	587	327	10
31° 00′	.5411	.5150	1.942	.6009	1.664	1.167	.8572	1.0297	59° 00′
10	440	175	932	048	653	169	557	268	50
20	469	200	923	088	643	171	542	239	40
30	.5498	.5225	1.914	.6128	1.632	1.173	.8526	1.0210	30
40	527	250	905	168	621	175	511	181	20
50	556	275	896	208	611	177	496	152	10
32° 00′	.5585	.5299	1.887	.6249	1.600	1.179	.8480	1.0123	58° 00′
10	614	324	878	289	590	181	465	094	50
20	643	348	870	330	580	184	450	065	40
30	.5672	.5373	1.861	.6371	1.570	1.186	.8434	1.0036	30
40	701	398	853	412	560	188	418	1.0007	20
50	730	422	844	453	550	190	403	977	10
33° 00′	.5760	.5446	1.836	.6494	1.540	1.192	.8387	.9948	57° 00′
10	789	471	828	536	530	195	371	919	50
20	818	495	820	577	520	197	355	890	40
30	.5847	.5519	1.812	.6619	1.511	1.199	.8339	.9861	30
40	876	544	804	661	501	202	323	832	20
50	905	568	796	703	1.492	204	307	803	10
34° 00′	.5934	.5592	1.788	.6745	1.483	1.206	.8290	.9774	56° 00′
		cos θ	sec θ	cot θ	tan θ	csc θ	sin θ	Radians	Degrees
								Angle θ	

TABLE I—*continued*

Angle θ Degrees	Radians	sin θ	csc θ	tan θ	cot θ	sec θ	cos θ		
34° 00′	.5934	.5592	1.788	.6745	1.483	1.206	.8290	.9774	56° 00′
10	963	616	781	787	473	209	274	745	50
20	992	640	773	830	464	211	258	716	40
30	.6021	.5664	1.766	.6873	1.455	1.213	.8241	.9687	30
40	050	688	758	916	446	216	225	657	20
50	080	712	751	959	437	218	208	628	10
35° 00′	.6109	.5736	1.743	.7002	1.428	1.221	.8192	.9599	55° 00′
10	138	760	736	046	419	223	175	570	50
20	167	783	729	089	411	226	158	541	40
30	.6196	.5807	1.722	.7133	1.402	1.228	.8141	.9512	30
40	225	831	715	177	393	231	124	483	20
50	254	854	708	221	385	233	107	454	10
36° 00′	.6283	.5878	1.701	.7265	1.376	1.236	.8090	.9425	54° 00′
10	312	901	695	310	368	239	073	396	50
20	341	925	688	355	360	241	056	367	40
30	.6370	.5948	1.681	.7400	1.351	1.244	.8039	.9338	30
40	400	972	675	445	343	247	021	308	20
50	429	995	668	490	335	249	004	279	10
37° 00′	.6458	.6018	1.662	.7536	1.327	1.252	.7986	.9250	53° 00′
10	487	041	655	581	319	255	969	221	50
20	516	065	649	627	311	258	951	192	40
30	.6545	.6088	1.643	.7673	1.303	1.260	.7934	.9163	30
40	574	111	636	720	295	263	916	134	20
50	603	134	630	766	288	266	898	105	10
38° 00′	.6632	.6157	1.624	.7813	1.280	1.269	.7880	.9076	52° 00′
10	661	180	618	860	272	272	862	047	50
20	690	202	612	907	265	275	844	.9018	40
30	.6720	.6225	1.606	.7954	1.257	1.278	.7826	.8988	30
40	749	248	601	.8002	250	281	808	959	20
50	778	271	595	050	242	284	790	930	10
39° 00′	.6807	.6293	1.589	.8098	1.235	1.287	.7771	.8901	51° 00′
10	836	316	583	146	228	290	753	872	50
20	865	338	578	195	220	293	735	843	40
30	.6894	.6361	1.572	.8243	1.213	1.296	.7716	.8814	30
40	923	383	567	292	206	299	698	785	20
50	952	406	561	342	199	302	679	756	10
40° 00′	.6981	.6428	1.556	.8391	1.192	1.305	.7660	.8727	50° 00′
10	.7010	450	550	441	185	309	642	698	50
20	039	472	545	491	178	312	623	668	40
30	.7069	.6494	1.540	.8541	1.171	1.315	.7604	.8639	30
40	098	517	535	591	164	318	585	610	20
50	127	539	529	642	157	322	566	581	10
41° 00′	.7156	.6561	1.524	.8693	1.150	1.325	.7547	.8552	49° 00′
		cos θ	sec θ	cot θ	tan θ	csc θ	sin θ	Radians	Degrees
								Angle θ	

158

TABLE I—*continued*

Angle θ		sin θ	csc θ	tan θ	cot θ	sec θ	cos θ		
Degrees	**Radians**								
41° 00′	.7156	.6561	1.524	.8693	1.150	1.325	.7547	.8552	**49° 00′**
10	185	583	519	744	144	328	528	523	50
20	214	604	514	796	137	332	509	494	40
30	.7243	.6626	1.509	.8847	1.130	1.335	.7490	.8465	30
40	272	648	504	899	124	339	470	436	20
50	301	670	499	952	117	342	451	407	10
42° 00′	.7330	.6691	1.494	.9004	1.111	1.346	.7431	.8378	**48° 00′**
10	359	713	490	057	104	349	412	348	50
20	389	734	485	110	098	353	392	319	40
30	.7418	.6756	1.480	.9163	1.091	1.356	.7373	.8290	30
40	447	777	476	217	085	360	353	261	20
50	476	799	471	271	079	364	333	232	10
43° 00′	.7505	.6820	1.466	.9325	1.072	1.367	.7314	.8203	**47° 00′**
10	534	841	462	380	066	371	294	174	50
20	563	862	457	435	060	375	274	145	40
30	.7592	.6884	1.453	.9490	1.054	1.379	.7254	.8116	30
40	621	905	448	545	048	382	234	087	20
50	650	926	444	601	042	386	214	058	10
44° 00′	.7679	.6947	1.440	.9657	1.036	1.390	.7193	.8029	**46° 00′**
10	709	967	435	713	030	394	173	.7999	50
20	738	988	431	770	024	398	153	970	40
30	.7767	.7009	1.427	.9827	1.018	1.402	.7133	.7941	30
40	796	030	423	884	012	406	112	912	20
50	825	050	418	942	006	410	092	883	10
45° 00′	.7854	.7071	1.414	1.000	1.000	1.414	.7071	.7854	**45° 00′**
		cos θ	sec θ	cot θ	tan θ	csc θ	sin θ	**Radians**	**Degrees**
								Angle θ	

TABLE II
Four-Place Values of Trigonometric Functions
Real Numbers *u*, or Angles θ, in Radians and Degrees

Real Number *u* or θ radians	θ degrees	sin *u* or sin θ	csc *u* or csc θ	tan *u* or tan θ	cot *u* or cot θ	sec *u* or sec θ	cos *u* or cos θ
0.00	0° 00′	0.0000	No value	0.0000	No value	1.000	1.000
.01	0° 34′	.0100	100.0	.0100	100.0	1.000	1.000
.02	1° 09′	.0200	50.00	.0200	49.99	1.000	0.9998
.03	1° 43′	.0300	33.34	.0300	33.32	1.000	0.9996
.04	2° 18′	.0400	25.01	.0400	24.99	1.001	0.9992
0.05	2° 52′	0.0500	20.01	0.0500	19.98	1.001	0.9988
.06	3° 26′	.0600	16.68	.0601	16.65	1.002	.9982
.07	4° 01′	.0699	14.30	.0701	14.26	1.002	.9976
.08	4° 35′	.0799	12.51	.0802	12.47	1.003	.9968
.09	5° 09′	.0899	11.13	.0902	11.08	1.004	.9960
0.10	5° 44′	0.0998	10.02	0.1003	9.967	1.005	0.9950
.11	6° 18′	.1098	9.109	.1104	9.054	1.006	.9940
.12	6° 53′	.1197	8.353	.1206	8.293	1.007	.9928
.13	7° 27′	.1296	7.714	.1307	7.649	1.009	.9916
.14	8° 01′	.1395	7.166	.1409	7.096	1.010	.9902
0.15	8° 36′	0.1494	6.692	0.1511	6.617	1.011	0.9888
.16	9° 10′	.1593	6.277	.1614	6.197	1.013	.9872
.17	9° 44′	.1692	5.911	.1717	5.826	1.015	.9856
.18	10° 19′	.1790	5.586	.1820	5.495	1.016	.9838
.19	10° 53′	.1889	5.295	.1923	5.200	1.018	.9820
0.20	11° 28′	0.1987	5.033	0.2027	4.933	1.020	0.9801
.21	12° 02′	.2085	4.797	.2131	4.692	1.022	.9780
.22	12° 36′	.2182	4.582	.2236	4.472	1.025	.9759
.23	13° 11′	.2280	4.386	.2341	4.271	1.027	.9737
.24	13° 45′	.2377	4.207	.2447	4.086	1.030	.9713
0.25	14° 19′	0.2474	4.042	0.2553	3.916	1.032	0.9689
.26	14° 54′	.2571	3.890	.2660	3.759	1.035	.9664
.27	15° 28′	.2667	3.749	.2768	3.613	1.038	.9638
.28	16° 03′	.2764	3.619	.2876	3.478	1.041	.9611
.29	16° 37′	.2860	3.497	.2984	3.351	1.044	.9582
0.30	17° 11′	0.2955	3.384	0.3093	3.233	1.047	0.9553
.31	17° 46′	.3051	3.278	.3203	3.122	1.050	.9523
.32	18° 20′	.3146	3.179	.3314	3.018	1.053	.9492
.33	18° 54′	.3240	3.086	.3425	2.920	1.057	.9460
.34	19° 29′	.3335	2.999	.3537	2.827	1.061	.9428
0.35	20° 03′	0.3429	2.916	0.3650	2.740	1.065	0.9394
Real Number *u* or θ radians	θ degrees	sin *u* or sin θ	csc *u* or csc θ	tan *u* or tan θ	cot *u* or cot θ	sec *u* or sec θ	cos *u* or cos θ

TABLE II—*continued*

Real Number *u* or θ radians	θ degrees	sin *u* or sin θ	csc *u* or csc θ	tan *u* or tan θ	cot *u* or cot θ	sec *u* or sec θ	cos *u* or cos θ
0.35	20° 03′	0.3429	2.916	0.3650	2.740	1.065	0.9394
.36	20° 38′	.3523	2.839	.3764	2.657	1.068	.9359
.37	21° 12′	.3616	2.765	.3879	2.578	1.073	.9323
.38	21° 46′	.3709	2.696	.3994	2.504	1.077	.9287
.39	22° 21′	.3802	2.630	.4111	2.433	1.081	.9249
0.40	22° 55′	0.3894	2.568	0.4228	2.365	1.086	0.9211
.41	23° 29′	.3986	2.509	.4346	2.301	1.090	.9171
.42	24° 04′	.4078	2.452	.4466	2.239	1.095	.9131
.43	24° 38′	.4169	2.399	.4586	2.180	1.100	.9090
.44	25° 13′	.4259	2.348	.4708	2.124	1.105	.9048
0.45	25° 47′	0.4350	2.299	0.4831	2.070	1.111	0.9004
.46	26° 21′	.4439	2.253	.4954	2.018	1.116	.8961
.47	26° 56′	.4529	2.208	.5080	1.969	1.122	.8916
.48	27° 30′	.4618	2.166	.5206	1.921	1.127	.8870
.49	28° 04′	.4706	2.125	.5334	1.875	1.133	.8823
0.50	28° 39′	0.4794	2.086	0.5463	1.830	1.139	0.8776
.51	29° 13′	.4882	2.048	.5594	1.788	1.146	.8727
.52	29° 48′	.4969	2.013	.5726	1.747	1.152	.8678
.53	30° 22′	.5055	1.978	.5859	1.707	1.159	.8628
.54	30° 56′	.5141	1.945	.5994	1.668	1.166	.8577
0.55	31° 31′	0.5227	1.913	0.6131	1.631	1.173	0.8525
.56	32° 05′	.5312	1.883	.6269	1.595	1.180	.8473
.57	32° 40′	.5396	1.853	.6410	1.560	1.188	.8419
.58	33° 14′	.5480	1.825	.6552	1.526	1.196	.8365
.59	33° 48′	.5564	1.797	.6696	1.494	1.203	.8309
0.60	34° 23′	0.5646	1.771	0.6841	1.462	1.212	0.8253
.61	34° 57′	.5729	1.746	.6989	1.431	1.220	.8196
.62	35° 31′	.5810	1.721	.7139	1.401	1.229	.8139
.63	36° 06′	.5891	1.697	.7291	1.372	1.238	.8080
.64	36° 40′	.5972	1.674	.7445	1.343	1.247	.8021
0.65	37° 15′	0.6052	1.652	0.7602	1.315	1.256	0.7961
.66	37° 49′	.6131	1.631	.7761	1.288	1.266	.7900
.67	38° 23′	.6210	1.610	.7923	1.262	1.276	.7838
.68	38° 58′	.6288	1.590	.8087	1.237	1.286	.7776
.69	39° 32′	.6365	1.571	.8253	1.212	1.297	.7712
0.70	40° 06′	0.6442	1.552	0.8423	1.187	1.307	0.7648
Real Number *u* or θ radians	θ degrees	sin *u* or sin θ	csc *u* or csc θ	tan *u* or tan θ	cot *u* or cot θ	sec *u* or sec θ	cos *u* or cos θ

TABLE II—*continued*

Real Number u or θ radians	θ degrees	sin u or sin θ	csc u or csc θ	tan u or tan θ	cot u or cot θ	sec u or sec θ	cos u or cos θ
0.70	40° 06′	0.6442	1.552	0.8423	1.187	1.307	0.7648
.71	40° 41′	.6518	1.534	.8595	1.163	1.319	.7584
.72	41° 15′	.6594	1.517	.8771	1.140	1.330	.7518
.73	41° 50′	.6669	1.500	.8949	1.117	1.342	.7452
.74	42° 24′	.6743	1.483	.9131	1.095	1.354	.7385
0.75	42° 58′	0.6816	1.467	0.9316	1.073	1.367	0.7317
.76	43° 33′	.6889	1.452	.9505	1.052	1.380	.7248
.77	44° 07′	.6961	1.436	.9697	1.031	1.393	.7179
.78	44° 41′	.7033	1.422	.9893	1.011	1.407	.7109
.79	45° 16′	.7104	1.408	1.009	.9908	1.421	.7038
0.80	45° 50′	0.7174	1.394	1.030	0.9712	1.435	0.6967
.81	46° 25′	.7243	1.381	1.050	.9520	1.450	.6895
.82	46° 59′	.7311	1.368	1.072	.9331	1.466	.6822
.83	47° 33′	.7379	1.355	1.093	.9146	1.482	.6749
.84	48° 08′	.7446	1.343	1.116	.8964	1.498	.6675
0.85	48° 42′	0.7513	1.331	1.138	0.8785	1.515	0.6600
.86	49° 16′	.7578	1.320	1.162	.8609	1.533	.6524
.87	49° 51′	.7643	1.308	1.185	.8437	1.551	.6448
.88	50° 25′	.7707	1.297	1.210	.8267	1.569	.6372
.89	51° 00′	.7771	1.287	1.235	.8100	1.589	.6294
0.90	51° 34′	0.7833	1.277	1.260	0.7936	1.609	0.6216
.91	52° 08′	.7895	1.267	1.286	.7774	1.629	.6137
.92	52° 43′	.7956	1.257	1.313	.7615	1.651	.6058
.93	53° 17′	.8016	1.247	1.341	.7458	1.673	.5978
.94	53° 51′	.8076	1.238	1.369	.7303	1.696	.5898
0.95	54° 26′	0.8134	1.229	1.398	0.7151	1.719	0.5817
.96	55° 00′	.8192	1.221	1.428	.7001	1.744	.5735
.97	55° 35′	.8249	1.212	1.459	.6853	1.769	.5653
.98	56° 09′	.8305	1.204	1.491	.6707	1.795	.5570
.99	56° 43′	.8360	1.196	1.524	.6563	1.823	.5487
1.00	57° 18′	0.8415	1.188	1.557	0.6421	1.851	0.5403
1.01	57° 52′	.8468	1.181	1.592	.6281	1.880	.5319
1.02	58° 27′	.8521	1.174	1.628	.6142	1.911	.5234
1.03	59° 01′	.8573	1.166	1.665	.6005	1.942	.5148
1.04	59° 35′	.8624	1.160	1.704	.5870	1.975	.5062
1.05	60° 10′	0.8674	1.153	1.743	0.5736	2.010	0.4976
Real Number u or θ radians	θ degrees	sin u or sin θ	csc u or csc θ	tan u or tan θ	cot u or cot θ	sec u or sec θ	cos u or cos θ

TABLE II—*continued*

Real Number u or θ radians	θ degrees	sin u or sin θ	csc u or csc θ	tan u or tan θ	cot u or cot θ	sec u or sec θ	cos u or cos θ
1.05	60° 10′	0.8674	1.153	1.743	0.5736	2.010	0.4976
1.06	60° 44′	8724	1.146	1.784	.5604	2.046	.4889
1.07	61° 18′	.8772	1.140	1.827	.5473	2.083	.4801
1.08	61° 53′	.8820	1.134	1.871	.5344	2.122	.4713
1.09	62° 27′	.8866	1.128	1.917	.5216	2.162	.4625
1.10	63° 02′	0.8912	1.122	1.965	0.5090	2.205	0.4536
1.11	63° 36′	.8957	1.116	2.014	.4964	2.249	.4447
1.12	64° 10′	.9001	1.111	2.066	.4840	2.295	.4357
1.13	64° 45′	.9044	1.106	2.120	.4718	2.344	.4267
1.14	65° 19′	.9086	1.101	2.176	.4596	2.395	.4176
1.15	65° 53′	0.9128	1.096	2.234	0.4475	2.448	0.4085
1.16	66° 28′	.9168	1.091	2.296	.4356	2.504	.3993
1.17	67° 02′	.9208	1.086	2.360	.4237	2.563	.3902
1.18	67° 37′	.9246	1.082	2.247	.4120	2.625	.3809
1.19	68° 11′	.9284	1.077	2.498	.4003	2.691	.3717
1.20	68° 45′	0.9320	1.073	2.572	0.3888	2.760	0.3624
1.21	69° 20′	.9356	1.069	2.650	.3773	2.833	.3530
1.22	69° 54′	.9391	1.065	2.733	.3659	2.910	.3436
1.23	70° 28′	.9425	1.061	2.820	.3546	2.992	.3342
1.24	71° 03′	.9458	1.057	2.912	.3434	3.079	.3248
1.25	71° 37′	0.9490	1.054	3.010	0.3323	3.171	0.3153
1.26	72° 12′	.9521	1.050	3.113	.3212	3.270	.3058
1.27	72° 46′	.9551	1.047	3.224	.3102	3.375	.2963
1.28	73° 20′	.9580	1.044	3.341	.2993	3.488	.2867
1.29	73° 55′	.9608	1.041	3.467	.2884	3.609	.2771
1.30	74° 29′	0.9636	1.038	3.602	0.2776	3.738	0.2675
1.31	75° 03′	.9662	1.035	3.747	.2669	3.878	.2579
1.32	75° 38′	.9687	1.032	3.903	.2562	4.029	.2482
1.33	76° 12′	.9711	1.030	4.072	.2456	4.193	.2385
1.34	76° 47′	.9735	1.027	4.256	.2350	4.372	.2288
1.35	77° 21′	0.9757	1.025	4.455	0.2245	4.566	0.2190
1.36	77° 55′	.9779	1.023	4.673	.2140	4.779	.2092
1.37	78° 30′	.9799	1.021	4.913	.2035	5.014	.1994
1.38	79° 04′	.9819	1.018	5.177	.1931	5.273	.1896
1.39	79° 38′	.9837	1.017	5.471	.1828	5.561	.1798
1.40	80° 13′	0.9854	1.015	5.798	0.1725	5.883	0.1700
Real Number u or θ radians	θ degrees	sin u or sin θ	csc u or csc θ	tan u or tan θ	cot u or cot θ	sec u or sec θ	cos u or cos θ

TABLE II—*continued*

Real Number u or θ radians	θ degrees	$\sin u$ or $\sin \theta$	$\csc u$ or $\csc \theta$	$\tan u$ or $\tan \theta$	$\cot u$ or $\cot \theta$	$\sec u$ or $\sec \theta$	$\cos u$ or $\cos \theta$
1.40	80° 13′	0.9854	1.015	5.798	0.1725	5.883	0.1700
1.41	80° 47′	.9871	1.013	6.165	.1622	6.246	.1601
1.42	81° 22′	.9887	1.011	6.581	.1519	6.657	.1502
1.43	81° 56′	.9901	1.010	7.055	.1417	7.126	.1403
1.44	82° 30′	.9915	1.009	7.602	.1315	7.667	.1304
1.45	83° 05′	0.9927	1.007	8.238	0.1214	8.299	0.1205
1.46	83° 39′	.9939	1.006	8.989	.1113	9.044	.1106
1.47	84° 13′	.9949	1.005	9.887	.1011	9.938	.1006
1.48	84° 48′	.9959	1.004	10.98	.0910	11.03	.0907
1.49	85° 22′	.9967	1.003	12.35	.0810	12.39	.0807
1.50	85° 57′	0.9975	1.003	14.10	0.0709	14.14	0.0707
1.51	86° 31′	.9982	1.002	16.43	.0609	16.46	.0608
1.52	87° 05′	.9987	1.001	19.67	.0508	19.69	.0508
1.53	87° 40′	.9992	1.001	24.50	.0408	24.52	.0408
1.54	88° 14′	.9995	1.000	32.46	.0308	32.48	.0308
1.55	88° 49′	0.9998	1.000	48.08	0.0208	48.09	0.0208
1.56	89° 23′	.9999	1.000	92.62	.0108	92.63	.0108
1.57	89° 57′	1.000	1.000	1256	.0008	1256	.0008
Real Number u or θ radians	θ degrees	$\sin u$ or $\sin \theta$	$\csc u$ or $\csc \theta$	$\tan u$ or $\tan \theta$	$\cot u$ or $\cot \theta$	$\sec u$ or $\sec \theta$	$\cos u$ or $\cos \theta$

TABLE III

Four-Place Logarithms of Numbers from 1 to 10
To extend the table write the number N as

$N = n \times 10^c, 1 \leq n < 10, c$ an integer, and use

$\log N = \log n + c.$

n	0	1	2	3	4	5	6	7	8	9
1.0	+0.0000	0043	0086	0128	0170	0212	0253	0294	0334	0374
1.1	.0414	0453	0492	0531	0569	0607	0645	0682	0719	0755
1.2	.0792	0828	0864	0899	0934	0969	1004	1038	1072	1106
1.3	.1139	1173	1206	1239	1271	1303	1335	1367	1399	1430
1.4	.1461	1492	1523	1553	1584	1614	1644	1673	1703	1732
1.5	.1761	1790	1818	1847	1875	1903	1931	1959	1987	2014
1.6	.2041	2068	2095	2122	2148	2175	2201	2227	2253	2279
1.7	.2304	2330	2355	2380	2405	2430	2455	2480	2504	2529
1.8	.2553	2577	2601	2625	2648	2672	2695	2718	2742	2765
1.9	.2788	2810	2833	2856	2878	2900	2923	2945	2967	2989
2.0	.3010	3032	3054	3075	3096	3118	3139	3160	3181	3201
2.1	.3222	3243	3263	3284	3304	3324	3345	3365	3385	3404
2.2	.3424	3444	3464	3483	3502	3522	3541	3560	3579	3598
2.3	.3617	3636	3655	3674	3692	3711	3729	3747	3766	3784
2.4	.3802	3820	3838	3856	3874	3892	3909	3927	3945	3962
2.5	.3979	3997	4014	4031	4048	4065	4082	4099	4116	4133
2.6	.4150	4166	4183	4200	4216	4232	4249	4265	4281	4298
2.7	.4314	4330	4346	4362	4378	4393	4409	4425	4440	4456
2.8	.4472	4487	4502	4518	4533	4548	4564	4579	4594	4609
2.9	.4624	4639	4654	4669	4683	4698	4713	4728	4742	4757
3.0	.4771	4786	4800	4814	4829	4843	4857	4871	4886	4900
3.1	.4914	4928	4942	4955	4969	4983	4997	5011	5024	5038
3.2	.5051	5065	5079	5092	5105	5119	5132	5145	5159	5172
3.3	.5185	5198	5211	5224	5237	5250	5263	5276	5289	5302
3.4	.5315	5328	5340	5353	5366	5378	5391	5403	5416	5428
3.5	.5441	5453	5465	5478	5490	5502	5514	5527	5539	5551
3.6	.5563	5575	5587	5599	5611	5623	5635	5647	5658	5670
3.7	.5682	5694	5705	5717	5729	5740	5752	5763	5775	5786
3.8	.5798	5809	5821	5832	5843	5855	5866	5877	5888	5899
3.9	.5911	5922	5933	5944	5955	5966	5977	5988	5999	6010
4.0	.6021	6031	6042	6053	6064	6075	6085	6096	6107	6117
4.1	.6128	6138	6149	6160	6170	6180	6191	6201	6212	6222
4.2	.6232	6243	6253	6263	6274	6284	6294	6304	6314	6325
4.3	.6335	6345	6355	6365	6375	6385	6395	6405	6415	6425
4.4	.6435	6444	6454	6464	6474	6484	6493	6503	6513	6522
4.5	.6532	6542	6551	6561	6571	6580	6590	6599	6609	6618
4.6	.6628	6637	6646	6656	6665	6675	6684	6693	6702	6712
4.7	.6721	6730	6739	6749	6758	6767	6776	6785	6794	6803
4.8	.6812	6821	6830	6839	6848	6857	6866	6875	6884	6893
4.9	.6902	6911	6920	6928	6937	6946	6955	6964	6972	6981

TABLE III—*continued*

n	0	1	2	3	4	5	6	7	8	9
5.0	+.6990	6998	7007	7016	7024	7033	7042	7050	7059	7067
5.1	.7076	7084	7093	7101	7110	7118	7126	7135	7143	7152
5.2	.7160	7168	7177	7185	7193	7202	7210	7218	7226	7235
5.3	.7243	7251	7259	7267	7275	7284	7292	7300	7308	7316
5.4	.7324	7332	7340	7348	7356	7364	7372	7380	7388	7396
5.5	.7404	7412	7419	7427	7435	7443	7451	7459	7466	7474
5.6	.7482	7490	7497	7505	7513	7520	7528	7536	7543	7551
5.7	.7559	7566	7574	7582	7589	7597	7604	7612	7619	7627
5.8	.7634	7642	7649	7657	7664	7672	7679	7686	7694	7701
5.9	.7709	7716	7723	7731	7738	7745	7752	7760	7767	7774
6.0	.7782	7789	7796	7803	7810	7818	7825	7832	7839	7846
6.1	.7853	7860	7868	7875	7882	7889	7896	7903	7910	7917
6.2	.7924	7931	7938	7945	7952	7959	7966	7973	7980	7987
6.3	.7993	8000	8007	8014	8021	8028	8035	8041	8048	8055
6.4	.8062	8069	8075	8082	8089	8096	8102	8109	8116	8122
6.5	.8129	8136	8142	8149	8156	8162	8169	8176	8182	8189
6.6	.8195	8202	8209	8215	8222	8228	8235	8241	8248	8254
6.7	.8261	8267	8274	8280	8287	8293	8299	8306	8312	8319
6.8	.8325	8331	8338	8344	8351	8357	8363	8370	8376	8382
6.9	.8388	8395	8401	8407	8414	8420	8426	8432	8439	8445
7.0	.8451	8457	8463	8470	8476	8482	8488	8494	8500	8506
7.1	.8513	8519	8525	8531	8537	8543	8549	8555	8561	8567
7.2	.8573	8579	8585	8591	8597	8603	8609	8615	8621	8627
7.3	.8633	8639	8645	8651	8657	8663	8669	8675	8681	8686
7.4	.8692	8698	8704	8710	8716	8722	8727	8733	8739	8745
7.5	.8751	8756	8762	8768	8774	8779	8785	8791	8797	8802
7.6	.8808	8814	8820	8825	8831	8837	8842	8848	8854	8859
7.7	.8865	8871	8876	8882	8887	8893	8899	8904	8910	8915
7.8	.8921	8927	8932	8938	8943	8949	8954	8960	8965	8971
7.9	.8976	8982	8987	8993	8998	9004	9009	9015	9020	9025
8.0	.9031	9036	9042	9047	9053	9058	9063	9069	9074	9079
8.1	.9085	9090	9096	9101	9106	9112	9117	9122	9128	9133
8.2	.9138	9143	9149	9154	9159	9165	9170	9175	9180	9186
8.3	.9191	9196	9201	9206	9212	9217	9222	9227	9232	9238
8.4	.9243	9248	9253	9258	9263	9269	9274	9279	9284	9289
8.5	.9294	9299	9304	9309	9315	9320	9325	9330	9335	9340
8.6	.9345	9350	9355	9360	9365	9370	9375	9380	9385	9390
8.7	.9395	9400	9405	9410	9415	9420	9425	9430	9435	9440
8.8	.9445	9450	9455	9460	9465	9469	9474	9479	9484	9489
8.9	.9494	9499	9504	9509	9513	9518	9523	9528	9533	9538
9.0	.9542	9547	9552	9557	9562	9566	9571	9576	9581	9586
9.1	.9590	9595	9600	9605	9609	9614	9619	9624	9628	9633
9.2	.9638	9643	9647	9652	9657	9661	9666	9671	9675	9680
9.3	.9685	9689	9694	9699	9703	9708	9713	9717	9722	9727
9.4	.9731	9736	9741	9745	9750	9754	9759	9763	9768	9773
9.5	.9777	9782	9786	9791	9795	9800	9805	9809	9814	9818
9.6	.9823	9827	9832	9836	9841	9845	9850	9854	9859	9863
9.7	.9868	9872	9877	9881	9886	9890	9894	9899	9903	9908
9.8	.9912	9917	9921	9926	9930	9934	9939	9943	9948	9952
9.9	.9956	9961	9965	9969	9974	9978	9983	9987	9991	9996

TABLE IV

Four-Place Logarithms of Trigonometric Functions
Angle θ in Degrees

Attach − 10 to Logarithms Obtained from This Table

Angle θ	L sin θ	L csc θ	L tan θ	L cot θ	L sec θ	L cos θ	
0° 00′	No value	No value	No value	No value	10.0000	10.0000	**90° 00′**
10′	7.4637	12.5363	7.4637	12.5363	.0000	.0000	50′
20′	.7648	.2352	.7648	.2352	.0000	.0000	40′
30′	7.9408	12.0592	7.9409	12.0591	.0000	.0000	30′
40′	8.0658	11.9342	8.0658	11.9342	.0000	.0000	20′
50′	.1627	.8373	.1627	.8373	.0000	10.0000	10′
1° 00′	8.2419	11.7581	8.2419	11.7581	10.0001	9.9999	**89° 00′**
10′	.3088	.6912	.3089	.6911	.0001	.9999	50′
20′	.3668	.6332	.3669	.6331	.0001	.9999	40′
30′	.4179	.5821	.4181	.5819	.0001	.9999	30′
40′	.4637	.5363	.4638	.5362	.0002	.9998	20′
50′	.5050	.4950	.5053	.4947	.0002	.9998	10′
2° 00′	8.5428	11.4572	8.5431	11.4569	10.0003	9.9997	**88° 00′**
10′	.5776	.4224	.5779	.4221	.0003	.9997	50′
20′	.6097	.3903	.6101	.3899	.0004	.9996	40′
30′	.6397	.3603	.6401	.3599	.0004	.9996	30′
40′	.6677	.3323	.6682	.3318	.0005	.9995	20′
50′	.6940	.3060	.6945	.3055	.0005	.9995	10′
3° 00′	8.7188	11.2812	8.7194	11.2806	10.0006	9.9994	**87° 00′**
10′	.7423	.2577	.7429	.2571	.0007	.9993	50′
20′	.7645	.2355	.7652	.2348	.0007	.9993	40′
30′	.7857	.2143	.7865	.2135	.0008	.9992	30′
40′	.8059	.1941	.8067	.1933	.0009	.9991	20′
50′	.8251	.1749	.8261	.1739	.0010	.9990	10′
4° 00′	8.8436	11.1564	8.8446	11.1554	10.0011	9.9989	**86° 00′**
10′	.8613	.1387	.8624	.1376	.0011	.9989	50′
20′	.8783	.1217	.8795	.1205	.0012	.9988	40′
30′	.8946	.1054	.8960	.1040	.0013	.9987	30′
40′	.9104	.0896	.9118	.0882	.0014	.9986	20′
50′	.9256	.0744	.9272	.0728	.0015	.9985	10′
5° 00′	8.9403	11.0597	8.9420	11.0580	10.0017	9.9983	**85° 00′**
10′	.9545	.0455	.9563	.0437	.0018	.9982	50′
20′	.9682	.0318	.9701	.0299	.0019	.9981	40′
30′	.9816	.0184	.9836	.0164	.0020	.9980	30′
40′	8.9945	11.0055	8.9966	11.0034	.0021	.9979	20′
50′	9.0070	10.9930	9.0093	10.9907	.0023	.9977	10′
6° 00′	9.0192	10.9808	9.0216	10.9784	10.0024	9.9976	**84° 00′**
	L cos θ	**L sec θ**	**L cot θ**	**L tan θ**	**L csc θ**	**L sin θ**	**Angle θ**

TABLE IV—*continued*
Attach − 10 to Logarithms Obtained from This Table

Angle θ	L sin θ	L csc θ	L tan θ	L cot θ	L sec θ	L cos θ	
6° 00′	9.0192	10.9808	9.0216	10.9784	10.0024	9.9976	**84° 00′**
10′	.0311	.9689	.0336	.9664	.0025	.9975	50′
20′	.0426	.9574	.0453	.9547	.0027	.9973	40′
30′	.0539	.9461	.0567	.9433	.0028	.9972	30′
40′	.0648	.9352	.0678	.9322	.0029	.9971	20′
50′	.0755	.9245	.0786	.9214	.0031	.9969	10′
7° 00′	9.0859	10.9141	9.0891	10.9109	10.0032	9.9968	**83° 00′**
10′	.0961	.9039	.0995	.9005	.0034	.9966	50′
20′	.1060	.8940	.1096	.8904	.0036	.9964	40′
30′	.1157	.8843	.1194	.8806	.0037	.9963	30′
40′	.1252	.8748	.1291	.8709	.0039	.9961	20′
50′	.1345	.8655	.1385	.8615	.0041	.9959	10′
8° 00′	9.1436	10.8564	9.1478	10.8522	10.0042	9.9958	**82° 00′**
10′	.1525	.8475	.1569	.8431	.0044	.9956	50′
20′	.1612	.8388	.1658	.8342	.0046	.9954	40′
30′	.1697	.8303	.1745	.8255	.0048	.9952	30′
40′	.1781	.8219	.1831	.8169	.0050	.9950	20′
50′	.1863	.8137	.1915	.8085	.0052	.9948	10′
9° 00′	9.1943	10.8057	9.1997	10.8003	10.0054	9.9946	**81° 00′**
10′	.2022	.7978	.2078	.7922	.0056	.9944	50′
20′	.2100	.7900	.2158	.7842	.0058	.9942	40′
30′	.2176	.7824	.2236	.7764	.0060	.9940	30′
40′	.2251	.7749	.2313	.7687	.0062	.9938	20′
50′	.2324	.7676	.2389	.7611	.0064	.9936	10′
10° 00′	9.2397	10.7603	9.2463	10.7537	10.0066	9.9934	**80° 00′**
10′	.2468	.7532	.2536	.7464	.0069	.9931	50′
20′	.2538	.7462	.2609	.7391	.0071	.9929	40′
30′	.2606	.7394	.2680	.7320	.0073	.9927	30′
40′	.2674	.7326	.2750	.7250	.0076	.9924	20′
50′	.2740	.7260	.2819	.7181	.0078	.9922	10′
11° 00′	9.2806	10.7194	9.2887	10.7113	10.0081	9.9919	**79° 00′**
10′	.2870	.7130	.2953	.7047	.0083	.9917	50′
20′	.2934	.7066	.3020	.6980	.0086	.9914	40′
30′	.2997	.7003	.3085	.6915	.0088	.9912	30′
40′	.3058	.6942	.3149	.6851	.0091	.9909	20′
50′	.3119	.6881	.3212	.6788	.0093	.9907	10′
12° 00′	9.3179	10.6821	9.3275	10.6725	10.0096	9.9904	**78° 00′**
10′	.3238	.6762	.3336	.6664	.0099	.9901	50′
20′	.3296	.6704	.3397	.6603	.0101	.9899	40′
30′	.3353	.6647	.3458	.6542	.0104	.9896	30′
40′	.3410	.6590	.3517	.6483	.0107	.9893	20′
50′	.3466	.6534	.3576	.6424	.0110	.9890	10′
13° 00′	9.3521	10.6479	9.3634	10.6366	10.0113	9.9887	**77° 00′**
	L cos θ	L sec θ	L cot θ	L tan θ	L csc θ	L sin θ	Angle θ

TABLE IV—*continued*

Attach — 10 to Logarithms Obtained from This Table

Angle θ	L sin θ	L csc θ	L tan θ	L cot θ	L sec θ	L cos θ	
13° 00′	9.3521	10.6479	9.3634	10.6366	10.0113	9.9887	**77° 00′**
10′	.3575	.6425	.3691	.6309	.0116	.9884	50′
20′	.3629	.6371	.3748	.6252	.0119	.9881	40′
30′	.3682	.6318	.3804	.6196	.0122	.9878	30′
40′	.3734	.6266	.3859	.6141	.0125	.9875	20′
50′	.3786	.6214	.3914	.6086	.0128	.9872	10′
14° 00′	9.3837	10.6163	9.3968	10.6032	10.0131	9.9869	**76° 00′**
10′	.3887	.6113	.4021	.5979	.0134	.9866	50′
20′	.3937	.6063	.4074	.5926	.0137	.9863	40′
30′	.3986	.6014	.4127	.5873	.0141	.9859	30′
40′	.4035	.5965	.4178	.5822	.0144	.9856	20′
50′	.4083	.5917	.4230	.5770	.0147	.9853	10′
15° 00′	9.4130	10.5870	9.4281	10.5719	10.0151	9.9849	**75° 00′**
10′	.4177	.5823	.4331	.5669	.0154	.9846	50′
20′	.4223	.5777	.4381	.5619	.0157	.9843	40′
30′	.4269	.5731	.4430	.5570	.0161	.9839	30′
40′	.4314	.5686	.4479	.5521	.0164	.9836	20′
50′	.4359	.5641	.4527	.5473	.0168	.9832	10′
16° 00′	9.4403	10.5597	9.4575	10.5425	10.0172	9.9828	**74° 00′**
10′	.4447	.5553	.4622	.5378	.0175	.9825	50′
20′	.4491	.5509	.4669	.5331	.0179	.9821	40′
30′	.4533	.5467	.4716	.5284	.0183	.9817	30′
40′	.4576	.5424	.4762	.5238	.0186	.9814	20′
50′	.4618	.5382	.4808	.5192	.0190	.9810	10′
17° 00′	9.4659	10.5341	9.4853	10.5147	10.0194	9.9806	**73° 00′**
10′	.4700	.5300	.4898	.5102	.0198	.9802	50′
20′	.4741	.5259	.4943	.5057	.0202	.9798	40′
30′	.4781	.5219	.4987	.5013	.0206	.9794	30′
40′	.4821	.5179	.5031	.4969	.0210	.9790	20′
50′	.4861	.5139	.5075	.4925	.0214	.9786	10′
18° 00′	9.4900	10.5100	9.5118	10.4882	10.0218	9.9782	**72° 00′**
10′	.4939	.5061	.5161	.4839	.0222	.9778	50′
20′	.4977	.5023	.5203	.4797	.0226	.9774	40′
30′	.5015	.4985	.5245	.4755	.0230	.9770	30′
40′	.5052	.4948	.5287	.4713	.0235	.9765	20′
50′	.5090	.4910	.5329	.4671	.0239	.9761	10′
19° 00′	9.5126	10.4874	9.5370	10.4630	10.0243	9.9757	**71° 00′**
10′	.5163	.4837	.5411	.4589	.0248	.9752	50′
20′	.5199	.4801	.5451	.4549	.0252	.9748	40′
30′	.5235	.4765	.5491	.4509	.0257	.9743	30′
40′	.5270	.4730	.5531	.4469	.0261	.9739	20′
50′	.5306	.4694	.5571	.4429	.0266	.9734	10′
20° 00′	9.5341	10.4659	9.5611	10.4389	10.0270	9.9730	**70° 00′**
	L cos θ	L sec θ	L cot θ	L tan θ	L csc θ	L sin θ	Angle θ

TABLE IV—*continued*
Attach −10 to Logarithms Obtained from This Table

Angle θ	L sin θ	L csc θ	L tan θ	L cot θ	L sec θ	L cos θ	
20 00'	9.5341	10.4659	9.5611	10.4389	10.0270	9.9730	**70° 00'**
10'	.5375	.4625	.5650	.4350	.0275	.9725	50'
20'	.5409	.4591	.5689	.4311	.0279	.9721	40'
30'	.5443	.4557	.5727	.4273	.0284	.9716	30'
40'	.5477	.4523	.5766	.4234	.0289	.9711	20'
50'	.5510	.4490	.5804	.4196	.0294	.9706	10'
21° 00'	9.5543	10.4457	9.5842	10.4158	10.0298	9.9702	**69° 00'**
10'	.5576	.4424	.5879	.4121	.0303	.9797	50'
20'	.5609	.4391	.5917	.4083	.0308	.9692	40'
30'	.5641	.4359	.5954	.4046	.0313	.9687	30'
40'	.5673	.4327	.5991	.4009	.0318	.9682	20'
50'	.5704	.4296	.6028	.3972	.0323	.9677	10'
22° 00'	9.5736	10.4264	9.6064	10.3936	10.0328	9.9672	**68° 00'**
10'	.5767	.4233	.6100	.3900	.0333	.9667	50'
20'	.5798	.4202	.6136	.3864	.0339	.9661	40'
30'	.5828	.4172	.6172	.3828	.0344	.9656	30'
40'	.5859	.4141	.6208	.3792	.0349	.9651	20'
50'	.5889	.4111	.6243	.3757	.0354	.9646	10'
23° 00'	9.5919	10.4081	9.6279	10.3721	10.0360	9.9640	**67° 00'**
10'	.5948	.4052	.6314	.3686	.0365	.9635	50'
20'	.5978	.4022	.6348	.3652	.0371	.9629	40'
30'	.6007	.3993	.6383	.3617	.0376	.9624	30'
40'	.6036	.3964	.6417	.3583	.0382	.9618	20'
50'	.6065	.3935	.6452	.3548	.0387	.9613	10'
24° 00'	9.6093	10.3907	9.6486	10.3514	10.0393	9.9607	**66° 00'**
10'	.6121	.3879	.6520	.3480	.0398	.9602	50'
20'	.6149	.3851	.6553	.3447	.0404	.9596	40'
30'	.6177	.3823	.6587	.3413	.0410	.9590	30'
40'	.6205	.3795	.6620	.3380	.0416	.9584	20'
50'	.6232	.3768	.6654	.3346	.0421	.9579	10'
25° 00'	9.6259	10.3741	9.6687	10.3313	10.0427	9.9573	**65° 00'**
10'	.6286	.3714	.6720	.3280	.0433	.9567	50'
20'	.6313	.3687	.6752	.3248	.0439	.9561	40'
30'	.6340	.3660	.6785	.3215	.0445	.9555	30'
40'	.6366	.3634	.6817	.3183	.0451	.9549	20'
50'	.6392	.3608	.6850	.3150	.0457	.9543	10'
26° 00'	9.6418	10.3582	9.6882	10.3118	10.0463	9.9537	**64° 00'**
10'	.6444	.3556	.6914	.3086	.0470	.9530	50'
20'	.6470	.3530	.6946	.3054	.0476	.9524	40'
30'	.6495	.3505	.6977	.3023	.0482	.9518	30'
40'	.6521	.3479	.7009	.2991	.0488	.9512	20'
50'	.6546	.3454	.7040	.2960	.0495	.9505	10'
27° 00'	9.6570	10.3430	9.7072	10.2928	10.0501	9.9499	**63° 00'**
	L cos θ	L sec θ	L cot θ	L tan θ	L csc θ	L sin θ	Angle θ

TABLE IV—*continued*
Attach − 10 to Logarithms Obtained from This Table

Angle θ	L sin θ	L csc θ	L tan θ	L cot θ	L sec θ	L cos θ	
27° 00′	9.6570	10.3430	9.7072	10.2928	10.0501	9.9499	**63° 00′**
10′	.6595	.3405	.7103	.2897	.0508	.9492	50′
20′	.6620	.3380	.7134	.2866	.0514	.9486	40′
30′	.6644	.3356	.7165	.2835	.0521	.9479	30′
40′	.6668	.3332	.7196	.2804	.0527	.9473	20′
50′	.6692	.3308	.7226	.2774	.0534	.9466	10′
28° 00′	9.6716	10.3284	9.7257	10.2743	10.0541	9.9459	**62° 00′**
10′	.6740	.3260	.7287	.2713	.0547	.9453	50′
20′	.6763	.3237	.7317	.2683	.0554	.9446	40′
30′	.6787	.3213	.7348	.2652	.0561	.9439	30′
40′	.6810	.3190	.7378	.2622	.0568	.9432	20′
50′	.6833	.3167	.7408	.2592	.0575	.9425	10′
29° 00′	9.6856	10.3144	9.7438	10.2562	10.0582	9.9418	**61° 00′**
10′	.6878	.3122	.7467	.2533	.0589	.9411	50′
20′	.6901	.3099	.7497	.2503	.0596	.9404	40′
30′	.6923	.3077	.7526	.2474	.0603	.9397	30′
40′	.6946	.3054	.7556	.2444	.0610	.9390	20′
50′	.6968	.3032	.7585	.2415	.0617	.9383	10′
30° 00′	9.6990	10.3010	9.7614	10.2386	10.0625	9.9375	**60° 00′**
10′	.7012	.2988	.7644	.2356	.0632	.9368	50′
20′	.7033	.2967	.7673	.2327	.0639	.9361	40′
30′	.7055	.2945	.7701	.2299	.0647	.9353	30′
40′	.7076	.2924	.7730	.2270	.0654	.9346	20′
50′	.7097	.2903	.7759	.2241	.0662	.9338	10′
31° 00′	9.7118	10.2882	9.7788	10.2212	10.0669	9.9331	**59° 00′**
10′	.7139	.2861	.7816	.2184	.0677	.9323	50′
20′	.7160	.2840	.7845	.2155	.0685	.9315	40′
30′	.7181	.2819	.7873	.2127	.0692	.9308	30′
40′	.7201	.2799	.7902	.2098	.0700	.9300	20′
50′	.7222	.2778	.7930	.2070	.0708	.9292	10′
32° 00′	9.7242	10.2758	9.7958	10.2042	10.0716	9.9284	**58° 00′**
10′	.7262	.2738	.7986	.2014	.0724	.9276	50′
20′	.7282	.2718	.8014	.1986	.0732	.9268	40′
30′	.7302	.2698	.8042	.1958	.0740	.9260	30′
40′	.7322	.2678	.8070	.1930	.0748	.9252	20′
50′	.7342	.2658	.8097	.1903	.0756	.9244	10′
33° 00′	9.7361	10.2639	9.8125	10.1875	10.0764	9.9236	**57° 00′**
10′	.7380	.2620	.8153	.1847	.0772	.9228	50′
20′	.7400	.2600	.8180	.1820	.0781	.9219	40′
30′	.7419	.2581	.8208	.1792	.0789	.9211	30′
40′	.7438	.2562	.8235	.1765	.0797	.9203	20′
50′	.7457	.2543	.8263	.1737	.0806	.9194	10′
34° 00′	9.7476	10.2524	9.8290	10.1710	10.0814	9.9186	**56° 00′**
	L cos θ	L sec θ	L cot θ	L tan θ	L csc θ	L sin θ	Angle θ

TABLE IV—*continued*
Attach − 10 to Logarithms Obtained from This Table

Angle θ	L sin θ	L csc θ	L tan θ	L cot θ	L sec θ	L cos θ	
34° 00′	9.7476	10.2524	9.8290	10.1710	10.0814	9.9186	**56° 00′**
10′	.7494	.2506	.8317	.1683	.0823	.9177	50′
20′	.7513	.2487	.8344	.1656	.0831	.9169	40′
30′	.7531	.2469	.8371	.1629	.0840	.9160	30′
40′	.7550	.2450	.8398	.1602	.0849	.9151	20′
50′	.7568	.2432	.8425	.1575	.0858	.9142	10′
35° 00′	9.7586	10.2414	9.8452	10.1548	10.0866	9.9134	**55° 00′**
10′	.7604	.2396	.8479	.1521	.0875	.9125	50′
20′	.7622	.2378	.8506	.1494	.0884	.9116	40′
30′	.7640	.2360	.8533	.1467	.0893	.9107	30′
40′	.7657	.2343	.8559	.1441	.0902	.9098	20′
50′	.7675	.2325	.8586	.1414	.0911	.9089	10′
36° 00′	9.7692	10.2308	9.8613	10.1387	10.0920	9.9080	**54° 00′**
10′	.7710	.2290	.8639	.1361	.0930	.9070	50′
20′	.7727	.2273	.8666	.1334	.0939	.9061	40′
30′	.7744	.2256	.8692	.1308	.0948	.9052	30′
40′	.7761	.2239	.8718	.1282	.0958	.9042	20′
50′	.7778	.2222	.8745	.1255	.0967	.9033	10′
37° 00′	9.7795	10.2205	9.8771	10.1229	10.0977	9.9023	**53° 00′**
10′	.7811	.2189	.8797	.1203	.0986	.9014	50′
20′	.7828	.2172	.8824	.1176	.0996	.9004	40′
30′	.7844	.2156	.8850	.1150	.1005	.8995	30′
40′	.7861	.2139	.8876	.1124	.1015	.8985	20′
50′	.7877	.2123	.8902	.1098	.1025	.8975	10′
38° 00′	9.7893	10.2107	9.8928	10.1072	10.1035	9.8965	**52° 00′**
10′	.7910	.2090	.8954	.1046	.1045	.8955	50′
20′	.7926	.2074	.8980	.1020	.1055	.8945	40′
30′	.7941	.2059	.9006	.0994	.1065	.8935	30′
40′	.7957	.2043	.9032	.0968	.1075	.8925	20′
50′	.7973	.2027	.9058	.0942	.1085	.8915	10′
39° 00′	9.7989	10.2011	9.9084	10.0916	10.1095	9.8905	**51° 00′**
10′	.8004	.1996	.9110	.0890	.1105	.8895	50′
20′	.8020	.1980	.9135	.0865	.1116	.8884	40′
30′	.8035	.1965	.9161	.0839	.1126	.8874	30′
40′	.8050	.1950	.9187	.0813	.1136	.8864	20′
50′	.8066	.1934	.9212	.0788	.1147	.8853	10′
40° 00′	9.8081	10.1919	9.9238	10.0762	10.1157	9.8843	**50° 00′**
10′	.8096	.1904	.9264	.0736	.1168	.8832	50′
20′	.8111	.1889	.9289	.0711	.1179	.8821	40′
30′	.8125	.1875	.9315	.0685	.1190	.8810	30′
40′	.8140	.1860	.9341	.0659	.1200	.8800	20′
50′	.8155	.1845	.9366	.0634	.1211	.8789	10′
41° 00′	9.8169	10.1831	9.9392	10.0608	10.1222	9.8778	**49° 00′**
	L cos θ	**L sec θ**	**L cot θ**	**L tan θ**	**L csc θ**	**L sin θ**	**Angle θ**

TABLE IV—*continued*

Attach − 10 to Logarithms Obtained from This Table

Angle θ	L sin θ	L csc θ	L tan θ	L cot θ	L sec θ	L cos θ	
41° 00′	9.8169	10.1831	9.9392	10.0608	10.1222	9.8778	**49° 00′**
10′	.8184	.1816	.9417	.0583	.1233	.8767	50′
20′	.8198	.1802	.9443	.0557	.1244	.8756	40′
30′	.8213	.1787	.9468	.0532	.1255	.8745	30′
40′	.8227	.1773	.9494	.0506	.1267	.8733	20′
50′	.8241	.1759	.9519	.0481	.1278	.8722	10′
42° 00′	9.8255	10.1745	9.9544	10.0456	10.1289	9.8711	**48° 00′**
10′	.8269	.1731	.9570	.0430	.1301	.8699	50′
20′	.8283	.1717	.9595	.0405	.1312	.8688	40′
30′	.8297	.1703	.9621	.0379	.1324	.8676	30′
40′	.8311	.1689	.9646	.0354	.1335	.8665	20′
50′	.8324	.1676	.9671	.0329	.1347	.8653	10′
43° 00′	9.8338	10.1662	9.9697	10.0303	10.1359	9.8641	**47° 00′**
10′	.8351	.1649	.9722	.0278	.1371	.8629	50′
20′	.8365	.1635	.9747	.0253	.1382	.8618	40′
30′	.8378	.1622	.9772	.0228	.1394	.8606	30′
40′	.8391	.1609	.9798	.0202	.1406	.8594	20′
50′	.8405	.1595	.9823	.0177	.1418	.8582	10′
44° 00′	9.8418	10.1582	9.9848	10.0152	10.1431	9.8569	**46° 00′**
10′	.8431	.1569	.9874	.0126	.1443	.8557	50′
20′	.8444	.1556	.9899	.0101	.1455	.8545	40′
30′	.8457	.1543	.9924	.0076	.1468	.8532	30′
40′	.8469	.1531	.9949	.0051	.1480	.8520	20′
50′	.8482	.1518	9.9975	.0025	.1493	.8507	10′
45° 00′	9.8495	10.1505	10.0000	10.0000	10.1505	9.8495	**45° 00′**
	L cos θ	L sec θ	L cot θ	L tan θ	L csc θ	L sin θ	Angle θ

ANSWERS TO ODD-NUMBERED PROBLEMS

Section 2. **1.** (a) $x > -1$; (c) $x < 8$; (e) $x > 5$; (g) $x = 0$; (i) $x > 3$ or $x < 2$; (k) $-1 \le x \le 0$ or $x \ge 1$; (m) $x \le -2$ or $0 \le x \le 1$; (o) $x < -1 - \sqrt{5}$ or $x > -1 + \sqrt{5}$; (q) $x < -2$ or $-1 < x < 1$; (s) $x < \frac{1}{2}$ or $x > \frac{4}{3}$.

Section 3. **1.** (a) $x = -5, 1$; (c) $x = 1, 1 + \sqrt{2}, 1 - \sqrt{2}$; (e) $-1 < x < 2$; (g) $x < -3$ or $x > 1$; (i) $x < -1$ or $x > 5$; (k) $x \le -\frac{1}{2}$.

Section 4. **1.** (a) $2\sqrt{17}$; (c) 13; (e) $\sqrt{37}$. **3.** (a) yes, $|AB| = \sqrt{65}$, $|AC| = 4$, $|BC| = 7$, $|AB|^2 = |AC|^2 + |BC|^2$; (c) No, $|AB| = 2\sqrt{13}$, $|AC| = \sqrt{29}$, $|BC| = \sqrt{37}$. **5.** (a) parallelogram; (c) parallelogram.

Section 6. **1.** (a) No, the pairs $(2,3), (2,5) \in f$, but $3 \ne 5$; (c) Yes, domain h = range $h = R$; (e) Yes, domain F = range $F = R$. **3.** $g(0) = \frac{3}{2}$, $g(3) = 0$, $g(-2) = \frac{5}{4}$, $g(1) = 2$, $g(4) = \frac{1}{2}$, domain $g = \{x : x \ne 2\}$. **5.** $f \div g = \{(x,a/b) : (x,a) \in f, (x,b) \in g, b \ne 0\}$.

Section 8. **1.** (a) $(5,3)$; (c) $(\frac{7}{6},\frac{1}{4})$; (e) $(5,-1)$; (g) $(-4,0)$.

Section 10. **3.** (a) $-\sin (\pi/5)$; (c) $-\tan (\pi/7)$; (e) $-\sec (\pi/15)$. **5.** The sine function is increasing for $0 < t < (\pi/2)$ and $(3\pi/2) < t < 2\pi$, and decreasing for $(\pi/2) < t < (3\pi/2)$. The cosine function is increasing for $\pi < t < 2\pi$, and decreasing for $0 < t < \pi$.

Section 11. 1. (a) $1/\sqrt{2}$; (c) $-1/\sqrt{3}$; (e) 2; (g) $-\sqrt{3}/2$; (i) 1. 3. (a) π; (c) $\pi/2$; (e) 2; (g) 1; (i) $\frac{4}{3}$.

Section 12. 3. $\sin t = 24/25$, $\tan t = -24/7$, $\cos t = -(7/24)$, $\sec t = -(25/7)$, $\csc t = 25/24$. 5. $\sin t = 5/\sqrt{74}$, $\cos t = 7/\sqrt{74}$.

Section 14. 3. $\sin(t+s) = -(63/65)$, $\cos(t-s) = -(56/65)$, $\tan(t+s) = 63/16$.

Section 15. 1. (a) neither even nor odd; (c) odd; (e) neither even nor odd; (g) neither even nor odd; (i) even. 5. (a) $(-x,y)$; (b) $(-x,-y)$.

Section 17. 1. l.u.b. cosine = 1, g.l.b. cosine = -1; $1 = \cos 0$ is the maximum, and $-1 = \cos \pi$ is the minimum. 3. l.u.b. $f = 4$, g.l.b. $f = 0$, $4 = f(2)$ is the maximum and $0 = f(0)$ is the minimum. 5. l.u.b. $f = 2$, g.l.b. $f = 0$, $2 = f(\pi/2)$ is the maximum and $0 = f(3\pi/2)$ is the minimum. 7. l.u.b. $h = 1$, g.l.b. $h = 0$, $1 = h(\pi/2)$ is the maximum and $0 = h(0)$ is the minimum. 9. No upper bound. g.l.b. $G = 0 = G(\pi/2)$, so also the minimum. 11. period = 4π, amplitude = ∞. 13. period = 2, amplitude = $\frac{1}{2}$.

Section 19. 1. period = $2\pi/3$, amplitude = 2, phase shift = 0. 3. period = 2, amplitude = $\frac{1}{2}$, phase shift = 0. 5. period = 4π, amplitude = 1, phase shift = π units to the right. 7. period = π, amplitude = $\frac{3}{4}$, phase shift = $\frac{3}{2}$ units to the left. 9. period = π, amplitude = 4, phase shift = $\frac{1}{2}$ unit to the right. 11. period = 2, amplitude = 3, phase shift = $1/6\pi$ units to the left. 13. period = $2\pi^2$, amplitude = 1, phase shift = $\pi/2$ units to the left. 15. period = $8\pi/3$, amplitude = 5, phase shift = $\pi/6$ units to the left.

Section 20. 1. period = π, amplitude = 2, phase shift = 0. 3. period = 4π, amplitude = 3, phase shift = 2 units to the left. 5. period = 2π, amplitude = ∞, phase shift = $\pi/4$ units to the right. 7. period = π, amplitude = ∞, phase shift = 0. 9. period = $\pi/2$, amplitude = ∞, phase shift = 0. 11. period = 2π, amplitude = ∞, phase shift = 0.

Section 21. 1. period = 2π, amplitude = $\sqrt{2}$, phase shift = $\pi/4$ units to the left. 3. period = π, amplitude = $\frac{1}{2}$, phase shift = 0. 5. The equation reduces to $y = 1$.

Section 22. 1. (a) $\pi/3$; (c) $\pi/2$; (e) $3\pi/4$; (g) $59\pi/90$; (i) $\pi/12$. 5. (a) $-\sin 57° = -0.8387$; (c) $-\tan 70° = -2.747$; (e) $-\cos 20° = -0.9397$. 7. $\cos(45° + 30°) = \sqrt{(2/4)}(\sqrt{3} - 1)$, $\cos(150°/2) = \sqrt{(1 + \cos 150°)/2} = \sqrt{(2 - \sqrt{3})/4} = \frac{1}{2}\sqrt{2 - \sqrt{3}}$. 9. (a) $(-1 + a)/(1 + a)$.

Section 23. 1. (a) $(1,\sqrt{3})$; (c) $(-4,0)$; (e) $(\frac{1}{3}, \sqrt{3}/3)$; (g) $(\sqrt{2}, -\sqrt{2})$. 3. (a) $[\sqrt{2}, \pi/4]$; (c) $[2\sqrt{5}, 63°]$; (e) $[3\sqrt{2}, 5\pi/4]$; (g) $[\sqrt{13}, 56°]$.

Section 25. 1. $c = 13$, $\alpha = 22°$, $\beta = 68°$. 3. $c = 5\sqrt{10}$, $\alpha = 55°$, $\beta = 35°$. 5. $\alpha = 25°$, $b = 21.4$, $c = 23.7$. 7. $\gamma = 105°$, $a = 67$, $b = 59$. 9. $\gamma = 70°$, $a = 53$, $b = 105$. 11. $\gamma = 75°$, $a = 73$, $b = 90$. 13. no solution. 15. $\beta = 26°$, $\gamma = 94°$, $c = 115$. 17. $\sqrt{2925}$ or approximately 54 feet. 19. 210 feet. 21. Height about 12,600 feet, total distance between 39,100 feet and 39,200 feet. 23. About 20°. 25. $|AC| = 297$ feet and $|BC| = 486$ feet.

Section 26. 1. $\alpha = 22°$, $\beta = 41°$, $\gamma = 107°$. 3. $\alpha = 26°$, $\beta = 63°$, $\gamma = 91°$. 5. $a = 19.7$, $\beta = 70°$, $\gamma = 43°$. 7. $b = 34.5$, $\alpha = 23°$, $\gamma = 45°$. 9. 20 inches. 11. The resultant is 53^+ lbs. and makes angles of 19° and 11° with the 20-lb. force and 35-lb. force, respectively.

Section 27. **1.** section **25: 1.** 30 sq. units. **3.** 58.5 sq. units. **5.** 107 sq. units. **7.** 1908 sq. units. **9.** 2619 sq. units. **11.** 17,572 sq. units. **13.** 2494 sq. units, section **26: 1.** $(625/4)\sqrt{231}$ sq. units; **3.** $(1875/4)\sqrt{455}$ sq. units; **5.** 138 sq. units; **7.** 181 sq. units.

Section 28. **1.** 5 inches. **3.** $\frac{3}{5}$ radians. **5.** 2.64 radians. **7.** $(1056/\pi)$ rpm. **9.** $20/3\pi$ feet. **11.** $v = 2000\pi/3$ feet/minute, angular velocity of blower = $2000/3$ rpm.

Section 29. **1.** Area of sector = $9\pi/2$ sq. inches, area of segment = $(9\pi/2) - 9\sqrt{2}$ sq. inches. **3.** Area of sector = $175\pi/24$ sq. inches, area of segment = $(175\pi/24) - (25\sqrt{2}/8)(\sqrt{3} + 1)$ sq. inches. **5.** $3\pi/2$ radians.

Section 30. **1.** (a) $f^c = \{(1,2), (4,3), (2,1), (1,5)\}$. f^{-1} does not exist; (c) $f^c = \{(x,y) : x = |y|, y \in \{0, 1, 2, 3, 4\}\} = \{(0,0), (1,1), (2,2), (3,3), (4,4)\}$, $f^c = f^{-1}$. **3.** $f^{-1} = \{(x,v) : y = (x + 2)/3\}$. **5.** $f^{-1} = \{(x,y) : y = \sqrt[3]{x + 1}\}$. **7.** $f^{-1} = \{(x,y) : y = (x - 3)/4\}$. **9.** No inverse $f(0) = f(-2) = 0$. **11.** $f^{-1} = \{(x,y) : y = \frac{1}{4}(-3 + \log_3 x)\}$. **13.** $f^{-1} = \{(x,y) : y = 5^{x - 1} - 2\}$. **15.** $f^{-1} = \{(x,y) : y = 2^{x - 1} - 3\}$. **17.** $f^{-1} = \{(x,y) : y = \frac{1}{2} - \log_5 2 + \log_5 \sqrt{x}\}$.

Section 32. **1.** (a) $\pi/6$; (c) $\pi/4$; (e) 0; (g) $\sqrt{3}/2$; (i) $2\sqrt{2}/3$.

Section 33. **1.** (a) $\pi/6$; (c) 0; (e) $\sqrt{3}$; (g) $2\pi/3$; (i) $1/\sqrt{2}$.

Section 34. **1.** $x = 0, \pi/6, \pi, 11\pi/6$. **3.** $x = \pi/3, 2\pi/3, 4\pi/3, 5\pi/3$. **5.** $x = \pi/3, \pi, 5\pi/3$. **7.** $x = 0$. **9.** $x = 0, \pi/3, 5\pi/3$. **11.** $x = \pi/7, 3\pi/7, 5\pi/7, \pi, 9\pi/7, 11\pi/7, 13\pi/7$. **13.** $x = 0, 2\pi/7, 4\pi/7, 6\pi/7, \pi, 8\pi/7, 10\pi/7, 12\pi/7$. **15.** $x = \pi/6, \pi/4, 3\pi/4, 5\pi/6, 5\pi/4, 3\pi/2, 7\pi/4$.

Section 37. **3.** (a) $-32i$; (c) $4 + 4i$; (e) $-8i$.

Section 38. **1.** $z_k = 2[\cos((5\pi/6) + k\pi) + i \sin((5\pi/6) + k\pi)]$, $z_0 = 2[\cos(5\pi/6) + i \sin(5\pi/6)] = -\sqrt{3} + i$, $z_1 = 2[\cos(11\pi/6) + i \sin(11\pi/6)] = \sqrt{3} - i$. **3.** $z_k = 2\sqrt[6]{2} \{\cos[(\pi/4) + (2k\pi/3)] + i \sin[(\pi/4) + (2k\pi/3)]\}$, $z_0 = 2\sqrt[6]{2}[\cos(\pi/4) + i \sin(\pi/4)] = \sqrt[6]{2}\sqrt{2}(1 + i)$, $z_1 = 2\sqrt[6]{2}[\cos(11\pi/12) + i \sin(11\pi/12)$, $z_2 = 2\sqrt[6]{2}[\cos(19\pi/12) + i \sin(19\pi/12)]$. **5.** $z_k = \sqrt{2}[\cos(0 + 2k\pi/6) + i \sin(0 + 2k\pi/6)]$, $z_0 = \sqrt{2}(\cos 0 + i \sin 0) = \sqrt{2}$, $z_1 = \sqrt{2}[\cos(\pi/3) + i \sin(\pi/3)] = (\sqrt{2}/2) + i(\sqrt{6}/2)$, $z_2 = \sqrt{2}[\cos(2\pi/3) + i \sin(2\pi/3)] = -(\sqrt{2}/2) + i(\sqrt{6}/2)$, $z_3 = \sqrt{2}[\cos \pi + i \sin \pi] = -\sqrt{2}$, $z_4 = \sqrt{2}[\cos(4\pi/3) + i \sin(4\pi/3)] = -\sqrt{2}/2 - i(\sqrt{6}/2)$, $z_5 = \sqrt{2}[\cos(5\pi/3) + i \sin(5\pi/3)] = (\sqrt{2}/2) - i(\sqrt{6}/2)$. **7.** $z_k = \cos[(\pi/8) + (k\pi/2)] + i \sin[(\pi/8) + (k\pi/2)]$, $z_0 = \cos(\pi/8) + i \sin(\pi/8)$, $z_1 = \cos(5\pi/8) + i \sin(5\pi/8)$, $z_2 = \cos(9\pi/8) + i \sin(9\pi/8)$, $z_3 = \cos(13\pi/8) + i \sin(13\pi/8)$. **9.** $z_k = \sqrt[3]{4}\{\cos[(5\pi/9) + (2k\pi/3)] + i \sin[(5\pi/9) + (2k\pi/3)]\}$; $z_0 = \sqrt[3]{4}[\cos(5\pi/9) + i \sin(5\pi/9)]$, $z_1 = \sqrt[3]{4}[\cos(11\pi/9) + i \sin(11\pi/9)]$, $z_2 = \sqrt[3]{4}[\cos(17\pi/9) + i \sin(17\pi/9)]$.

Appendix A. **1.** (a) 1; (c) 3; (e) 0; (g) 0; (i) 0; (l) -4. **3.** (a) 0.7781; (c) $8.3801 - 10$; (e) $9.6990 - 10$. **5.** (a) 3.81; (c) -34.9; (e) 1.69; (g) 1540.

Appendix B. **1.** $\alpha = 81°$, $\beta = 55°$, $c = 16.5$. **3.** $\alpha = 58°$, $\gamma = 82°$, $b = 0.095$. **5.** $\alpha = 46°$, $\beta = 56°$, $\gamma = 78°$.

INDEX